Synthesis Lectures on Renewable Energy Technologies

The series, Synthesis Lectures on Renewable Energy Technologies publishes concise books, focused on technologies that harness energy from naturally occurring sources, such as sunlight, wind, water, geothermal heat, and biofuels from organic materials. These renewable energy technologies play a crucial role in transitioning away from fossil fuels, helping to mitigate the effects of climate change, and promoting a sustainable energy supply.

Neyre Tekbıyık Ersoy

Energy Efficiency and Renewable Energy Policies

Springer

Neyre Tekbıyık Ersoy
Energy Systems Engineering
Faculty of Engineering
Cyprus International University
Nicosia, North Cyprus via Mersin 10, Türkiye

ISSN 2690-5000 ISSN 2690-5019 (electronic)
Synthesis Lectures on Renewable Energy Technologies
ISBN 978-3-031-64307-1 ISBN 978-3-031-64305-7 (eBook)
https://doi.org/10.1007/978-3-031-64305-7

This Springer imprint is published by the registered company Springer Nature Switzerland AG
The registered company address is: Gewerbestrasse 11, 6330 Cham, Switzerland

Dedicated to my family

Preface

I have been teaching and making research about energy policies for many years now. I remember the time that I first needed a textbook that my students can easily follow and learn about energy laws and policies. Although there were many publications about such policies, most of them were reports, articles, journals, or such other publications. There were also many other specific publications (either books or reports, etc.) focusing on regional or specific types of energy efficiency policies or renewable energy policies, but not a textbook prepared in the way that I needed. As I usually have students coming from different locations around the World, I wanted to teach them the common nature of those policies, not just the policies focusing on one region or category. This way, they could understand the reasoning behind those policies and how their application changes from one country to another. This would help them to adopt those policies to their countries/regions if needed, or simply to have the skills in order to develop new policies when needed. Thus, I started making more research and collecting more information from many different resources to construct my own lecture notes and update them whenever necessary. The more I updated my lecture notes over the years, the more generalized they have become. At one stage, I decided that I want to share this knowledge not just with my own students, but with everyone who wants to teach and learn about energy efficiency policies and renewable energy policies. This is how the book idea has arisen. I wanted this book to include detailed information about each one of the policies but at the same time, to be easy to understand and follow as it was going to be a textbook. I wanted the readers to be able to navigate their way through the book to simply reach the parts that they want to read as soon as possible. Hence, this book has been organized in a way that it can serve the above-mentioned purposes.

This book provides an in-depth but easy to navigate analysis of the energy efficiency and renewable energy policies. It starts with the definition of energy efficiency and the need for it, and elaborates on the types of energy efficiency policies used in different sectors; building sector, transport sector, industrial sector, and power sector. Then, it provides a comprehensive analysis about which countries have more energy efficiency policies in which sectors. After completion of energy efficiency policies, the book continues with renewable energy and why it is needed. Then, it provides descriptions and characteristics of different types of renewable energy policies and guides the reader about developing a successful renewable energy policy. The book continues with sector-based analysis of renewable energy (RE) policies; RE policies in heating and cooling sector, RE policies in transport sector, and RE policies in power sector. Finally, the book provides a comprehensive analysis about which countries are supporting which RE technologies.

Although this book is designed to be used as a textbook, it can also be attractive to those consumers willing to learn more about the energy efficiency policies that can help them to lower down their consumption and thus the utility bills. The book is also useful for those willing to learn more about renewable energy and how it is supported via different measures. Hence, this book appeals to a wide audience; from policymakers to decisionmakers, from producers to consumers, and finally from teachers to students.

Nicosia, North Cyprus, Türkiye — Neyre Tekbıyık Ersoy

Contents

Abbreviations

BEAP	Business Energy Advice Program
BRT	Bus Rapid Transit
CFL	Compact Fluorescent Light Bulb
CHP	Combined Heat and Power
CSP	Concentrated Solar Power
DSM	Demand Side Management
DTP	Dynamic Tidal Power
EDC	Energy Development Corporation
EE	Energy Efficiency
EEPS	Energy Efficiency Portfolio Standards
EERS	Energy Efficiency Resource Standards
EPA	Environmental Protection Agency
ERDF	European Regional Development Fund
ESCO	Energy Service Company
EU	European Union
EV	Electric Vehicle
FIP	Feed-in Premium
FIT	Feed-in Tariff
GACC	Global Alliance for Clean Cookstoves
GDP	Gross Domestic Product
GDRE	General Directorate of Renewable Energy
GHG	Greenhouse Gas
GW	Gigawatt
HAWT	Horizontal Axis Wind Turbine
HVAC	Heating, Ventilation, and Air Conditioning
IEA	International Energy Agency
IPCC	Intergovernmental Panel on Climate Change
IRENA	International Renewable Energy Agency
ITC	Investment Tax Credit
ITS	Intelligent Traffic or Transport Systems

kW	Kilowatt
Lao PDR	Lao People's Democratic Republic
LCFS	Low Carbon Fuel Standard
LED	Light-Emitting Diode
LEZ	Low Emission Zone
LPG	Liquid Petroleum Gas
LPO	Loan Programs Office
MELS	Minimum Energy Labeling Schemes
MEPS	Minimum Energy Performance Standards
MINVU	Ministry of Housing and Urban Development
MOE	Ministry of Electricity of the Republic of Iraq
MPG	Miles Per Gallon Of Fuel
MW	Megawatt
MWh	Megawatt hour
NIMBY	Not in My Backyard
OECD	Organization for Economic Co-Operation and Development
P2P	Peer-To-Peer
ppm	Parts Per Million
PTC	Production Tax Credit
PV	Photovoltaic
R&D	Research and Development
RE	Renewable Energy
REAP	Rural Energy for America Program
REC	Renewable Energy Certificate
REPI	Renewable Energy Production Incentive
RHI	Renewable Heat Incentive
RPS	Renewable Portfolio Standard
SDG	Sustainable Development Goal
SMEs	Small to Medium Enterprises
TEN-T	Trans-European Transport Network
TOU	Time of Use
U.S.	United States
UK	United Kingdom
UN	United Nations
URF	Uganda Road Fund
USA	United States of America
VAT	Value Added Tax
VAWT	Vertical Axis Wind Turbine
ZEZ	Zero Emission Zone

1 Energy Efficiency and the Need for Energy Efficiency

Energy efficiency and energy conservation are often complimentary or sometimes overlapping ways to reduce energy consumption. Energy efficiency (EE) refers to the ability of a system to decrease its energy consumption while providing the same level of service. It is related with the technical performance of energy conversion and energy-consuming devices [1]. Energy conservation, on the other hand, covers actions to reduce the end user related energy consumption. For example, installing energy-efficient light bulbs (such as Compact fluorescent light bulbs (CFLs), or Light-emitting diodes (LEDs)) is an efficiency measure, while turning lights off when not in need is a conservation measure. Energy efficiency helps the consumers to decrease their consumption and thus pay less bills. According to [2], some of the energy efficiency examples can be listed as follows:

- Boiler replacement
  - Replacing conventional boilers with high-efficiency models
- Building automation
  - Controlling the building's air-conditioning, heating, ventilation, and lighting systems in a centralized manner, with building automation systems
- Cogeneration
  - Using waste heat from electricity generation for heating purposes
- District heating/cooling
  - Replacing the individual (or building based) heating and cooling systems with large-scale district heating and cooling systems
- Economizer on existing boilers
  - Capturing residual heat in the boiler system and using it to preheat the boiler input

N. Tekbıyık Ersoy, *Energy Efficiency and Renewable Energy Policies*, Synthesis Lectures on Renewable Energy Technologies, https://doi.org/10.1007/978-3-031-64305-7_1

- Heat pump replacing boiler
  - Extraction and transfer of heat from outside air, water or soil via heat pumps
- Improvement of illumination
  - Using modern illumination systems that require less energy for the same luminous flux or adopting occupancy sensors, etc.
- Installation of thermostatic valves
  - Regulating the water flow through the radiator via thermostatic valves
- Optimization of compressed air systems
  - Utilizing energy efficient compressed air systems to reduce energy losses
- Replacement of air conditioning systems
  - Switching to efficient air conditioning systems
- Sealing and exchange of windows
  - Reducing the heat loss via sealing leaks in windows or using double or triple glazed windows
- Thermal insulation of roof and walls
  - Better insulation for a roof or the walls
- Variable speed drives
  - Achieving higher efficiency levels via controlling the speed of electric motors
- Ventilation
  - Reducing energy needs of ventilation systems through better ventilation controls
- Waste heat recovery
  - Using the waste heat arising from industrial processes for the purposes of space, air or water heating

The interested reader is referred to [2] for more information about such applications and possible energy savings. As already mentioned above, energy efficiency measures help people reduce their individual energy consumption. But at the same time, it protects them from possible increase in energy costs. This can be explained as follows; when the demand increases, the supply should also increase. This means that government or the utility company should invest in capacity additions or system improvements. Considering that the energy generation and supply costs increase, this in turn causes an increase in the energy costs paid by the consumers. Energy efficiency is usually considered as the cheapest, fastest, and most reliable new energy resource. Usually, energy efficiency programs do not require new power lines or pipelines, and emit almost no air pollutants. Although EE provides direct benefits, energy efficient appliances or systems may not always be adopted/deployed as required, due to some obstacles. The next subsection explains those obstacles and provides possible ways that can be used in overcoming those obstacles.

1.1 Obstacles for Deployment of Energy Efficiency

According to [3], the obstacles for energy efficiency deployment are; lack of awareness, lack of knowledge and skills, higher up-front investments, and split incentives. Especially in the last few decades, energy demand has grown dramatically. As the balance between demand and supply is the key to energy security, this increases the need for more energy generation and therefore the related investments. Many people seeing the price of electricity increasing with time want to reduce their energy consumption for different reasons; to consume less and therefore pay less, or may be to consume less and save the environment. However, unfortunately, many of those people have little awareness about individual opportunities for saving energy. For example; a house owner who is using electrical water boiler, may not be aware of the feasibility of using a solar water heater (to substitute the electrical onc) in their region. Similarly some architects or engineers may not be aware of how their design would significantly affect the heating and cooling demand of the building [3]. Furthermore, in some regions of the world, although people know that they can save energy with the usage of certain technologies, they may not have the knowledge about how to use or adopt that technology to their region. Even if they want to buy and apply that technology to their house, they may not be able to access related labor force (engineering, technician, etc.) with necessary skills.

When buying electrical appliances many people face with a trade-off. An appliance that is more efficient versus an appliance that is cheaper. Mostly, due to economic reasons, they prefer to choose the cheaper one. However, when life-cycle costs of the same appliances are compared, it is revealed that they may end up paying more rather than paying less. Another dilemma is caused by split incentives. Split incentives arise when those who purchase equipment do not have to pay for energy operating costs [4]. Most home developers and landlords make decisions based on initial capital costs. This is due to the fact that if the owner of the building makes an investment to improve the energy efficiency of the building, the tenant is the one to get the resulting financial savings from reduced energy bills. Therefore, in a real estate market where there is not much of a competition, a house owner willing to get more profit would have no incentive to invest in EE technologies.

1.2 How to Remove Those Obstacles

The above-mentioned obstacles can be overcome with the following measures [3]:

- Awareness campaigns that show people ways of saving energy in their house
- Training courses for architects, engineers, technicians on energy conservation, energy efficiency and sustainability

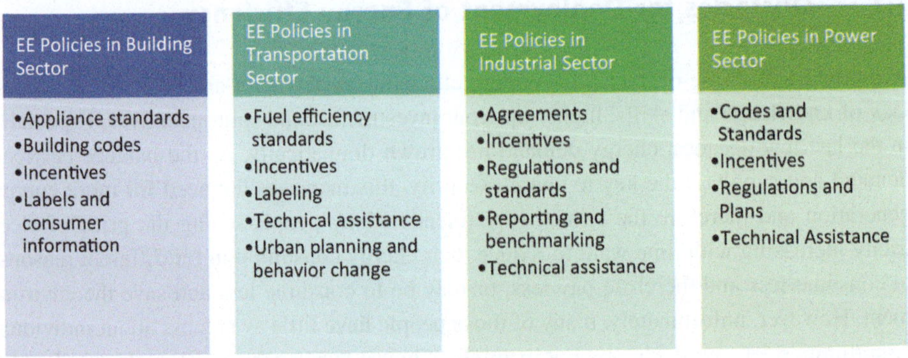

Fig. 1.1 Types of energy efficiency policies based on sectors

- Regulations that allow the landlord or the house owner to add the extra costs of the efficiency investments to the rent

But, selecting the most appropriate energy efficiency support policies for a country or a region can be the best solution for removing any kind of obstacle that may arise. The next sections are devoted to the energy efficiency policies applied in various sectors: Building sector, transportation sector, industrial sector and power sector. The types of policies applied in each sector are listed in Fig. 1.1. These policies will be elaborated in the following chapters.

1.3 Exercises

1. What is the difference between energy efficiency and energy conservation?
2. Which type of energy efficiency technologies or methods can be applied in buildings?
3. Which energy efficiency applications can reduce the heating related energy demand?
4. What are the obstacles for deployment of energy efficiency?
5. What are split incentives? How do they affect energy efficiency adoption?
6. How can the obstacles related with energy efficiency be removed?
7. What are the types of energy efficiency policies applied in building sector?
8. What are the types of energy efficiency policies applied in transport sector?
9. What are the types of energy efficiency policies applied in industrial sector?
10. What are the types of energy efficiency policies applied in power sector?

References

1. EIA. (2024). *Use of energy explained: Energy efficiency and conservation.* U.S. Energy Information Administration. Retrieved April 2024, from https://www.eia.gov/energyexplained/use-of-energy/efficiency-and-conservation.php
2. Agster, R., Hector, S., Hennig, C., & Schürmann, S. (2019). *Energy efficiency manual an overview of energy efficiency measures that are targeted by the PF4EE instrument*. European Investment Bank. Adelphi Consult Gmbh. Retrieved April 2024, from https://pf4ee.eib.org/sites/default/files/2019-09/PF4EE-ESF_SEP2019_EE%20Manual.pdf
3. Langniss, O. (2010). *Promoting sustainable energy in the Bahamas.* Fichtner. Retrieved April 2010, from http://www.best.gov.bs/Webdocs/1016_FinalReport.pdf
4. Doris, E., Cochran, J., & Vorum, M. (2009). *Energy efficiency policy in the United States: Overview of trends at different levels of government.* National Renewable Energy Laboratory. Retrieved April 2024, from https://www.nrel.gov/docs/fy10osti/46532.pdf

Energy Efficiency Policies in Building Sector 2

The building sector includes energy used for constructing, heating, cooling and lighting houses and businesses, as well as the appliances and equipment installed in these buildings. According to UN Environment programme [1], this sector accounts for approximately 37% of energy and process-related CO_2 emissions and more than 34% of energy demand globally. Hence, lowering down the energy consumption in this sector would significantly help the countries in terms of balancing the demand and supply and freeing up the financial resources. Reducing the energy consumption requires changing daily choices in energy consumption and adopting energy efficient technologies. Many different policies can be applied in building sector in order to increase the use of energy-efficient technologies. This section will focus on four major policy types mostly employed in building sector: Appliance standards, building codes, incentives and, labels and consumer information.

2.1 Appliance Standards

Appliance standards set the minimum energy efficiency requirements for the sale of specific products. They are often called minimum energy performance standards (MEPS). In order to set the standards, governments usually consider a consensus and negotiation of standards that the industry can meet with reasonable increase in prices. Usually, before going into a mandatory phase, a voluntary target is applied. Then, as the market transformation continues, the targets are introduced as standards [2]. Depending on the type of

N. Tekbıyık Ersoy, *Energy Efficiency and Renewable Energy Policies*, Synthesis Lectures on Renewable Energy Technologies, https://doi.org/10.1007/978-3-031-64305-7_2

the appliance standard, it may target both residential appliances (e.g. televisions, showerheads, etc.) and commercial ones (e.g. dishwashers, lighting, etc.). The standards may target different performance metrics, such as [3]:

- The amount of energy used when an appliance is in standby mode
- The maximum rate that water comes out of a faucet/showerhead
- The amount of cooling power required per watt-hour of electricity

Appliance standards may apply to products manufactured, sold or imported into the country or region. Hence, manufacturers can produce more efficient appliances than the minimum, however, the appliance standards remove the option for consumers to purchase inefficient alternatives. Especially, industry-wide standards are helpful in reducing adoption costs and the cost of efficiency. Because, in order to compete in the price-sensitive part of the market, the manufacturers try to find the least-cost ways of reducing energy consumption [4]. Appliance standards also reduce the effects of various barriers to the long term energy savings. These barriers are listed by 4Cleanair [5] as follows:

- Lack of consumer awareness on benefits of energy efficient products
- Lack of information on energy efficient products
- Financial procedures overemphasizing initial costs and de-emphasizing operating costs
- Limited stock of energy efficient products
- Manufacturer price competition
- Split incentives between building owners and renters (house owner's demotivation for buying energy efficient appliances, as the house will be used by the renter)

4Cleanair [5] claims that appliance standards save tremendous amount of energy and the associated emissions from power plants, at the lowest possible cost. However, different levels of stringency of appliance standards may cause different outcomes. Too strong standards may lead to over-investment in energy efficiency, causing a burden for both manufacturers and consumers. The burden mentioned here is due to products being more expensive than the amount people will recover from their utility bill savings. In contrary to that, if standards are too weak, low-quality products become widespread, contributing to higher electricity bills [2].

2.2 Building Codes

A building code is a set of regulations (written in mandatory, enforceable language) governing the design, construction, alteration, and maintenance of structures and equipment. It is also called building energy standard, and specifies how a building/equipment must be constructed or perform [6]. Building codes regulate the certain aspects of a building envelope, lighting, and heating, ventilation, and air conditioning (HVAC) system in order to help occupants to save energy and money over a building's lifetime [4]. The most widely used building energy standards and codes are; separate energy efficiency requirements (prescriptive code), and energy performance requirements (performance code). Prescriptive codes are set for each component of the building (such as thermal transfer values for walls, roof, windows, etc.) and for each part of the equipment (heating/cooling system, lights, pumps, etc.) [7]. Performance codes, on the other hand, are based on annual energy consumption or the building's greenhouse gas emissions.

They may vary across the World and show regional differences due to varying climates of different regions. However, the differences are not only caused by climate. The ways of implementing the building codes can also vary significantly. Some countries may prefer to first develop a plan for implementation of building codes in a single sector or for certain types of buildings (such as larger buildings) in order to test the code and learn from their experience. On the other hand, some countries may start by adopting building codes for new buildings, as a legal obligation for construction approval. This is due to the fact that implementing energy efficient technologies during construction of a new building is often less costly than retrofitting an existing inefficient building. However, building codes can also be applied in existing buildings. Some countries, such as Germany, apply the building codes to existing buildings when they are renovated. In Germany, if more than 20% of the building area is to be renovated, a certain level of energy performance is required for renovations [7].

As described above, countries may start with a limited scope of building code implementation, but it is important to have a longer term roadmap for implementation of building codes across the whole building sector [8]. Upon implementation, building codes should be monitored, evaluated and improved over time. Because, unfortunately, compliance and enforcement of building codes is the key challenge. Even in developed countries, compliance is not easy due to the high transaction costs required for inspection and verification. According to [7], in the United Kingdom, only 40% of new buildings comply with the building codes, while compliance in the Netherlands is as low as 20% due to reluctance to enforce regulations on building owners. The non-compliance can be due to building owners who are more sensitive to construction costs rather than energy related costs over the life cycle of buildings. Another reason could be lack of finance as energy efficient technologies are usually more expensive than the others. There can also be cases where the energy efficient equipment is not available in the local market. This happens mostly in developing countries, and requires imports of the related technologies, making the costs higher than they already are.

Local governments are usually the most critical actors in supporting effective compliance and building code enforcement. In order to provide a successful implementation strategy, in case of non-compliance to building codes, penalties should be applied. Possible penalties for not complying with the building code can include stopping construction, withholding permits and levying fines. Also, as technology progresses and the costs decline, building codes should be regularly updated to remain relevant and still be effective. Various stages of building code implementation and enforcement can be found in [8]. When building codes are applied properly, the tenants, who live in buildings compliant with building codes, can save on their utility bills due to the energy savings achieved from the installation of technologies that meet energy-efficiency requirements. According to [9], today's building codes provide 30% more energy savings than the codes from a decade ago, and the residential/commercial building codes are expected to save $126 billion in cumulative energy costs between 2010 and 2040.

2.3 Incentives

Energy efficient technologies can have high capital costs relative to inefficient alternatives. In order to help reduce this cost barrier and improve development, some policies may offer financial incentives. The types of financial incentives are as follows: Grants and subsidies, loans, rebates, and tax incentives. Grants and subsidies aim to defray the costs of investment, while loans aim to improve access to credit for investment in new systems, technologies, or efficiency measures. In rebates however, the buyer purchases an appliance or system that satisfies specific use and energy efficiency requirements, and later receives a rebate. Rebates can be considered as the most common type of incentive, as they offer a cost-effective way for utilities to reduce the energy demand. Tax incentives can be in the form of credits, rebates, or exemptions to offset investment costs for energy efficiency measures. Among these options, tax credit stands out as its application differs from other options. Tax-credit refers to the amount of money that taxpayers can subtract from taxes upon the purchase of certain energy efficiency appliances, or investment/installation of an energy efficiency system. For example, homeowners who make approved energy efficiency improvements to their houses (such as better insulation, windows, heating and cooling systems, etc.) can subtract a percentage of these upgrade costs up to a certain credit limit (determined by the government).

Different countries may apply the above-mentioned types of incentives differently. For example, according to [10], the Estonian Ministry of Economy offers energy efficiency subsidies for apartment buildings retrofits. Based on the location and the size of the building, the subsidies can cover 30–50% of costs of complete retrofits of apartment buildings with energy efficiency class C or lower. Similarly, in Canada, the Federal Government provides the Greener Homes Grant to help homeowners increase their home's energy efficiency. Up to 700,000 grants of up to CAD 5,000 are available for energy efficient retrofits. Eligible upgrades can be listed as; replacing windows/doors, sealing air leaks, adding insulation, improving heating and cooling systems, etc. In Latvia, on the other hand, the Latvian development finance institution Altum provides state loans for energy efficiency renovations of apartment buildings. Eligible projects can receive up to 400,000 euros per apartment building and up to 3 million euros per landlord.

Incentives are considered as an effective way to overcome market barriers, attract customer attention, and encourage homeowners to invest in energy assessments and energy efficiency related upgrades. Incentives can be financial or non-financial. Although financial incentives might improve interest in a program, usually, funding for such incentives is limited. Moreover, although high incentive amounts can attract the consumers, their success depends on the design of the program. For those interested in designing an effective incentive program, Better Buildings Residential Network [11] provides a range of tools to help programs; identify successful incentives for residential energy efficiency programs, learn to implement incentives, and assess incentive effectiveness.

2.4 Labeling and Education

Labeling conveys information on energy operating costs to help consumers make purchasing decisions. This is done as follows: Manufacturer mandatorily discloses the energy consumption of appliances with labels. Although some customers may not have prior knowledge about energy related issues, when they want to buy new appliances, they see that labels on the appliances and naturally start comparing them with other such appliances. This increases awareness about energy consumption. Moreover, the customer is unconsciously educated about which type of devices consume more electricity. Actually, according to [12], in an EU-wide (Eurobarometer) survey in 2019, 93% of the consumers confirmed recognizing the label and 79% confirmed that it had influenced their decision on which product to buy.

Labeling may have different forms, such as comparative labels and endorsements. Comparative labels inform the consumers about the annual energy consumption of a product relative to other products in the same class. Endorsements, on the other hand, certify (via a symbol placed on a product) that the product is one of the most energy efficient in its class. The application of labeling can also differ across the countries or regions. For example, the European Union (EU) has issued a directive for labelling of washing machines, tumble dryers, dishwashers, refrigerators, electric ovens, air conditioners, and lamps. EU Member states have the responsibility of ensuring that the labelling schemes are accompanied by educational and promotional information campaigns, in order to encourage more responsible use of energy by the consumers [13].

According to [4], one drawback of labeling programs is their focus on efficiency rather than the total consumption. As products are categorized by classes, this allows large, energy-consuming devices to be a separate category from similar products with smaller energy consumption. This enables them to offer more energy-consuming features without losing energy-efficiency endorsements. Hence, a cap on total energy consumption of energy efficient products may be required. Few years ago, with more and more products achieving ratings as A+, A++ or A+++, a shift to new rating scale became necessary. The shift happened in 2021. This new scale is stricter and designed in a way that only few products are initially able to achieve the "A" rating [12]. This leaves a space for more efficient products to be included in the future. When the shift to new rating system happened, the most energy efficient products of the time received the following labels; "B", "C" or "D". Hence, since March 2021, the energy label rating system uses A to G rankings, being applied to the following product groups; refrigerators, dishwashers, washing machines, televisions, light bulbs and lamps [14]. A and G indicate the most and the least energy efficient devices accordingly. European Commission [12] explains the differences between the old scale and the new scale for EU energy labels. For example, for a fridge, the new label shows not only the energy consumption, but also many other details about the fridge, such as; capacity, noise level, etc. More detailed information about building energy standards and labelling in Europe can be found in [15].

It should be noted that energy labels are not special to appliances. Buildings can also have energy labels. Actually, energy-labelling of buildings is mandatory in some countries, such as Denmark [16]. Building energy label aims to disclose the energy consumption of the building and the potential energy-saving measures. When a building is energy-labelled, it is inspected and measured by an energy consultant. The energy consultant calculates the building's energy consumption. The energy label is based on a theoretical calculation of energy consumption under some assumptions about user behavior and weather conditions. Therefore, it indicates the energy standard of the building, rather than the building's actual consumption. This way, it provides a basis for comparing the state and quality of different buildings. The labels classify buildings on a scale from A to G. A covers low energy buildings, which only consume a minimum amount of energy, while G-labelled buildings consume the most energy [17]. In addition to labeling, information campaigns can be used in order to raise awareness. However, in order to do that, first, tools (such as questionnaires, or data analysis) should be utilized for assessing the energy demand, and awareness level of the consumers.

2.5 Exercises

1. Which type of barriers to the long term energy savings can be reduced by appliance standards?
2. What are the major differences between prescriptive codes and performance codes?
3. What are the possible reasons of non-compliance to building codes? What can be done to reduce non-compliance?
4. What are the types of financial incentives applied in building sector in order to support energy efficiency?
5. What is tax-credit? What makes it different than other types of financial incentives?

6. What are the possible benefits of using labeling (over others) to promote energy efficiency in building sector?
7. Since 2021, which ratings indicate the most and the least energy efficient devices?
8. What is the main goal of building energy label?

References

1. UNEP. (2024). *Sustainable buildings.* UN Environment Programme. Retrieved April 2024, from https://www.unep.org/topics/cities/buildings-and-construction/sustainable-buildings
2. UNESCAP. (2024). *Appliance standards and labelling. Low carbon green growth roadmap for Asia and the Pacific: Fact sheet.* Retrieved April 2024, from https://www.unescap.org/sites/default/d8files/2021-11/1.%20FS-Appliance-standards-and-labelling.pdf
3. Climate Xchange Staff. (2023). *Policy explainer: How appliance standards can cut emissions and increase savings.* Climate Xchange. Retrieved April 2024, from https://climate-xchange.org/2023/09/08/policy-explainer-how-appliance-standards-can-cut-emissions-and-increase-savings/
4. Doris, E., Cochran, J., & Vorum, M. (2009). *Energy efficiency policy in the United States: Overview of trends at different levels of government.* National Renewable Energy Laboratory. Retrieved April 2024, from https://www.nrel.gov/docs/fy10osti/46532.pdf
5. 4Cleanair. (2024). *Chapter 14. Boost appliance efficiency standards. Implementing EPA's clean power plan: A menu of options.* Retrieved April 2024, from https://www.4cleanair.org/wp-content/uploads/Documents/Chapter_14.pdf
6. Rowan, L. R. (2023). *Building codes, standards, and regulations: Frequently asked questions.* Congressional Research Service. Retrieved April 2024, from https://sgp.fas.org/crs/misc/R47665.pdf
7. UNESCAP. (2024). *Building energy standards and codes. Low carbon green growth roadmap for Asia and the Pacific: Fact sheet.* Retrieved April 2024, from https://www.unescap.org/sites/default/d8files/2021-11/4.%20FS-Building-energy-standards-and-codes.pdf
8. Cox, S. (2016). *Building energy codes policy overview and good practices.* National Renewable Energy Laboratory. Clean Energy Solutions Center. Retrieved April 2024, from https://www.nrel.gov/docs/fy16osti/65542.pdf
9. Myers, A. (2020). *Building codes: A powerful yet underused climate policy that could save billions.* Forbes. Retrieved April 2024, from https://www.forbes.com/sites/energyinnovation/2020/12/02/a-powerful-yet-underused-climate-tool-building-codes/
10. IEA. (2024). *Policies database.* IEA. Retrieved April 2024, from https://www.iea.org/policies
11. Better Buildings Residential Network. (2024). *Designing incentives toolkit.* US Department of Energy. Retrieved April 2024, from https://www.energy.gov/eere/better-buildings-residential-network/articles/voluntary-initiative-designing-incentives
12. European Commission. (2024). *New EU energy labels applicable from 1 March 2021.* European Commission. Retrieved April 2024, from https://ec.europa.eu/commission/presscorner/detail/en/ip_21_818
13. OECD. (2008). *Promoting sustainable consumption: Good practices in OECD countries.* OECD Publications. Retrieved April 2024, from https://www.oecd.org/greengrowth/40317373.pdf

14. Directorate General for Internal Market, Industry, Entrepreneurship and SMES. (2024). *Energy label*. European Union. Retrieved April 2024, from https://europa.eu/youreurope/business/product-requirements/labels-markings/energy-labels/index_en.htm
15. Moore, C., Shrestha, S., & Gokarakonda, S. (2019). *Building energy standards and labelling in Europe*. Wuppertal Institute. Retrieved April 2024, from https://www.switch-asia.eu/site/assets/files/2287/building_energy_standards_and_labelling_in_europe.pdf
16. Østergård, T. (2019). *Energy labels and energy efficient properties: Methods for identification and definition of energy efficient properties in Denmark*. MOE. Retrieved April 2024, from https://www.nykredit.com/siteassets/ir/files/bond-issuance/green-bonds/moe_report_energy_labels_and_energy_efficient_properties_2019-01-25.pdf
17. The Danish Energy Agency. (2024). *Energy labels for buildings*. Danish Energy Agency. Retrieved April 2024, from https://ens.dk/en/our-responsibilities/energy-labels-buildings

3 Energy Efficiency Policies in Transportation Sector

According to International Energy Agency (IEA) [1], as the end of 2021, the energy consumption in transportation sector constitutes almost 27% of total final consumption and transportation sector related emissions constitute almost 24% of CO_2 emissions, globally. Evidently, transportation sector has a significant effect on the World's energy demand and increasing emission levels. The relationship between energy and transport is direct but may vary based on different transport modes, such as; air, road, rail and marine. Passengers and high-value goods are usually transported by fast and energy-intensive modes, due to the high value of the time component of their mobility. The maritime transportation, on the other hand, is linked to low energy consumption per unit of mass transported, with a slower speed. This is appropriate for freight transport, as in that case, time is less critical and stocks can be accumulated. Contrarily, air freight has high energy consumption levels due to the high-speed services with limited stocks [2].

Improving transportation efficiencies can address the movement of people or goods. For example, for the movement of people, the focus can be on reducing the fuel needed per mile driven, reducing the total distance driven, encouraging the use of public transport, etc. For the movement of goods, on the other hand, focusing on increasing intermodal transportation could offer significant benefits. Intermodal transportation refers to moving goods in the steel-based containers through two or more modes of transport. This method reduces cargo handling, improves security, reduces loss, and allows goods to be transported faster. As it may be understood from the discussions above, energy efficiency policies in transportation sector focus more on developing new technologies that increase fuel efficiency and incentivize changing transportation patterns. Transportation

N. Tekbıyık Ersoy, *Energy Efficiency and Renewable Energy Policies*, Synthesis Lectures on Renewable Energy Technologies, https://doi.org/10.1007/978-3-031-64305-7_3

sector related energy efficiency policies can be categorized as follows: fuel efficiency standards, incentives, labeling, technical assistance, and urban planning and behavior change.

3.1 Fuel Efficiency Standards

Fuel efficiency standards aim to make vehicles more efficient so that they consume less fuel. This can be achieved by improving existing technology so that petroleum and diesel fueled vehicles operate more efficiently. The standard specifies the average fuel efficiency (km/ℓ) that vehicle manufacturers should follow. The standard may also provide an obligation on light vehicle suppliers to make sure that new vehicles entering into the market, meet a particular CO_2/km standard. Manufacturers are liable for fines/penalties if their vehicles do not meet the required standards.

Different countries may apply different fuel efficiency standards. For example, according to [3], in Switzerland, the importers of light commercial vehicles have to limit the average of the imported fleet to 147 gCO_2/km from 2020. In China, the 2019 Stage III National Standard tightens fuel consumption limits for new tractors, trucks and buses by 15.3, 13.8 and 15.9% compared to the Stage II Standard. These standards promote technological innovation in fuel efficiency, and are important tools in reducing long-term fuel demand.

3.2 Incentives

Incentives in transportation sector can be financial or non-financial. The main aim of such incentives is to encourage manufactures to develop fuel-efficient technologies and consumers to purchase that technologies. Financial incentives can be in the form of low-interest loans and consumer-oriented financial incentives [4]. Low-interest loans appeal to manufacturers to upgrade their manufacturing process and/or to produce fuel-efficient equipment. Consumer-oriented financial incentives, on the other hand, concentrate on lowering the upfront cost of fuel-efficient technologies.

Incentives do not always have to be financial. Actually, in cases where financial incentives are undesirable (due to reduced government revenues or opposition to higher taxes on traditional cars), non-financial incentives may be more effective. For example, access to traffic-restricted roads or lower-traffic, high-occupancy lanes. This could also be in the form of access to preferential parking. Especially in urban areas with high amount of traffic, time spent on traffic can be much higher than desired. Also, finding a place for a car can be problematic. Therefore, such non-financial incentives can have a really motivating nature. According to [5], many major Latin American cities already have programs to restrict car use in order to reduce traffic and air pollution.

3.3 Labeling and Consumer Education

Fuel-consumption labels inform consumers about the fuel efficiency and potential operating costs of a vehicle. Such labels provide fuel efficiency estimates for cities and highway driving in an easy-to-read sticker placed prominently on new vehicles for sale [4]. According to [6], in European Union (EU), an energy label includes the following information about the vehicle: brand/model, fuel type, weight, fuel usages (in litre/100 km), average CO_2 emissions of the full test cycle (in g/km), etc. At this stage, it should be noted that even in EU, the member states implement the labelling systems differently. The United States (U.S.) energy labelling system focuses on fuel economy instead of energy usage. The vehicle's fuel performance is indicated by miles per gallon of fuel (MPG). Their label includes many different parameters as seen in Fig. 3.1 [7].

In the figure, (1) shows the vehicle technology and fuel (such as diesel vehicle, electric vehicle, etc.). (2) shows the combined fuel economy as a weighted average of City and Highway MPG values (weighting the City value by 55% and the Highway value by 45%). (3) compares the fuel economy to the other vehicles. The category of the vehicle (For example, Small SUV, Station Wagon, etc.) and the worst and best fuel economy within that category can be seen there. (4) shows the estimated fuel cost over a 5-year period for the vehicle, compared to an average new vehicle. (5) expresses the fuel efficiency in terms of consumption (gallons per mile or per 100 miles). The estimated annual

Fig. 3.1 The U.S. energy labelling system focusing on fuel economy [7]

fuel cost is shown in (6). As seen in the figure, the fuel economy and greenhouse gas rating (from 1 (worst) to 10 (best)) is shown in (7). (8) enhances that information by stating combined city/highway CO_2 tailpipe emissions and vehicle with lowest CO_2 emissions. (9) completes the environmental information by stating the rating (1 (worst) to 10 (best)) for vehicle tailpipe emissions of those pollutants that cause smog and other air pollution. (10) reminds that the fuel economy and emissions may be different due to the many factors, such as; the way that the vehicle is driven or maintained, the use of air conditioning and other accessories, the weather, road conditions, etc. The QR Code® in (11) links the consumer to helpful tools and additional information about the vehicle. Finally, (12) directs the consumer to the fueleconomy.gov website, where they can compare vehicles and enter personalized information in order to obtain the best possible cost and energy-use estimates. Labels slightly differ for plug-in hybrids and electric cars. Interested reader is referred to [7] for more information about different labels in U.S., and to [8, 9] for different types of fuel consumption labels applied in various countries. No matter how the labels look, they provide significant information about the fuel and energy efficiency of a vehicle. This educates the consumers without them realizing it, and helps them to make budget friendly and environmentally friendly choices.

3.4 Technical Assistance

Technical assistance programs draw on the expertise of government agencies to guide municipalities and businesses on how to improve energy efficiency. In the U.S., various programs exist about this issue. For example; Alternative Fuels Data Center and Clean Cities Technical Assistance. Alternative fuels data center, which can be accessed via [10], includes many tools that can assist fuel providers, fleets and other decision makers about how to advance energy-efficient vehicle technologies [11]. Among the many tools available in that data center, AFLEET tool can be used to calculate a fleet's cost of ownership, petroleum use and emissions. ATRAVEL tool can be used to estimate travel time, costs and emissions for private vehicles and other travel modes. Alternative Fueling Station Locator can be used to locate alternative fueling stations and get maps and driving directions. The other technical assistance program, Clean Cities Technical Assistance, includes the Technical Response Service. This service can help people in finding answers to their technical questions about fuel economy improvements, idle reduction measures, alternative fuels and advanced vehicles [10]. More information about the Clean Technical Assistance program, such as; the outreach materials and templates for idle reduction projects and electric vehicle infrastructure projects, and educational resources for school districts implementing electric school buses, can be found in [12].

Technical assistance programs may change significantly from one country to another. For example, since 2017, in Argentina, trainings were held in many different cities, such as; Rosario, Posadas, Paraná, Cipolletti, etc. The topics addressed in the trainings can be listed as follows: efficient driving, efficient fleet management, control and monitoring of fuel consumption, technology for savings applied to transport and cost impact. IEA [3] claims that more than 330 entrepreneurs and 50 presidents of provincial and sectoral chambers were trained as a result of this system. In the same year, 2017, the European Union started providing technical assistance to the Uganda Road Fund (URF), within the scope of far reaching institutional reforms in the transportation sector. The technical assistance to URF was designed to improve its institutional capacity and corporate governance, and thus, the operational efficiency of road maintenance [13].

3.5 Urban Planning and Behavior Change

Most probably, one of the most effective low-cost and long-term measures to reduce fuel consumption is to change the consumer behavior. This can be done via various measures, such as urban planning (prioritizing walking and public transportation), zoning, traffic design and idle reduction.

3.5.1 Urban Planning

People usually spend a lot of time while driving from home to school or work (and to return back to home) every day. Considering the other daily activities that a person may require, such as visits to shopping center, health center, sports center, etc., indicates that city planning has a considerable effect on person's daily transport energy consumption. Keeping this in mind, urban planning related policies consider the design of a city or a region in a way that the people will need to drive less and prefer walking or public transportation. As a good example to this, Inturri and Ignaccolo [14] lists some of the principles that have a clear impact on transport energy savings and (Greenhouse gas) GHG reduction as follows:

- Providing financial incentives to encourage the residents to live near where they work
- Encouraging developers to reduce off-street surface parking as it undermines the walkability that compact communities could otherwise support
- Setting the right price for street parking in order to motivate the residents to use less cars
- Concentrating major services near homes, jobs and public transport
- Ensuring easy access to public transport by encouraging higher-density residential development around stops
- Providing bike racks at public transport stations in order to make public transport systems more appealing (as every transit trip starts and ends with walking)

Böhler-Baedeker and Hüging [15] provides case studies about some local authorities from Portland and San Francisco which are limiting the maximum parking capacity at particular sites or within a particular area. The main goal is to discourage the use of cars and promote the use of public transport. Plate restriction schemes can also be used for forcing car users to switch to more efficient modes of transport. For example, Böhler-Baedeker and Hüging [15] refers to an example application in Bogotá in which 40% of private vehicles cannot operate in the city different time periods within the day (7:00–9:00 and 17:30–19:30), in accordance with designated number plates. A similar application exists in Sao Paulo.

In addition to the schemes mentioned above, policies that would make cycling more attractive within urban environment can also be successful. An example to that is the promotion of bicycle lanes and cycle highways. Although using bicycle could be preferred by some people, they may hesitate due to safety concerns on the roads. Bicycle lanes and cycle highways could offer solutions to this problem. According to [3], the federal government of Germany adopted a National Cycling Plan, designed to initiate new strategies and improvements for promoting cycling up to 2012. Within that scope, in 2008, the government is reported to invest about €100 m for the construction and maintenance of cycling paths on trunk roads and in cycling safety work. A similar example is seen in Czech Republic, the 2nd pillar of the Czech National Recovery Plan. This plan is concerned with physical infrastructures and green transition including several measures such as intermodal transport, digitization of railway infrastructure, cycle paths, etc. As of June 1, 2021, the allocated budget is reported to be CZK 24 billion excluding VAT [3].

Some countries are also making investments on cycle highways (or Bicycle Highways), which are paved cycling routes that are separated from regular motor vehicle traffic. The main goal of such highways are to increase efficiency and reduce travel times. A good application example can be seen in Belgium, the Flemish cycle highway network [16] [Aleksander Buczyński/European Cyclists' Federation (analysis) and OpenStreetMap (background map)]. This network is a comprehensive plan for connecting cities/towns, suburbs and workplaces across the whole region via the high-quality cycling infrastructure. Its scale can easily be seen in Fig. 3.2.

In order to decrease the fuel usage, public transportation should also be encouraged. Hence, policies focusing on that would have a significant impact on energy reduction in transportation sector. Bus rapid transit system (BRT) can be considered as a good example [15]. BRT system is a high-capacity, low-cost public transit solution that uses buses on dedicated lanes in order to quickly and efficiently transport passengers to their destination. It can attract choice riders and greatly increase corridor ridership. NBRTI [17] claims that ridership gains of 20–96% in BRT corridors have been noted in practice. For a real life application example, mainly the TransMilenio BRT System, the interested reader is referred to [18].

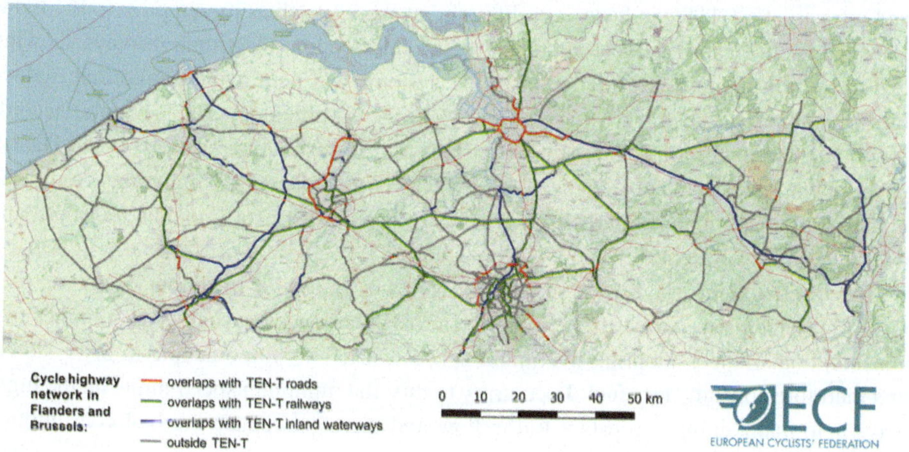

Fig. 3.2 Map showing crossover between Flemish cycle highway network and (Trans-European Transport Network) TEN-T infrastructure [16]

3.5.2 Zoning

Zoning is the process of dividing a region into different areas having specific rules and regulations about land use, design and development. There can be many different zoning applications from one region to another. For example, many European cities have vehicle entry regulations depending on the types or emissions of the vehicles, payments, etc. Within the scope of energy efficiency policies in transport sector, two zoning applications will be described; environmental zones and pedestrian zones. Environmental zones are areas which only vehicles that meet a prescribed emission standard are allowed to enter. Usually, such zones are designed to improve the regional air quality. However, in some cases, this requirement can also be used to encourage more energy-efficient vehicles [15]. Pedestrian zones, on the other hand, are based on designating some core city areas as pedestrian zones to discourage people from using their personal vehicles.

3.5.3 Traffic Design/Traffic Planning

Traffic planning focuses on evaluating the current traffic trends in a region and seeking for solutions in order to streamline traffic for the future. Traffic planning related schemes reducing energy consumption in transportation sector include, but are not limited to; intelligent traffic or transport systems (ITS) and speed restrictions. Intelligent transport systems include a range of technology, software and physical infrastructure that make travelling around cities more efficient. ITS technologies consist of sensing, communication, and data analytics to monitor/control different elements of transportation systems; vehicles, roads, and infrastructure. ITS can have very different measures, such as pre-emptive traffic lights activated by the transit vehicle approaching the intersection, electronic displays showing real-time information, etc. Intelligent traffic systems can help avoid congestion, reduce fuel consumption and travel time and improve energy efficiency. For example, according to [3], the long-term Sustainable Transport and Infrastructure strategy of the Danish government has several objectives. One of them is to reduce CO_2 emissions from road transport by combining; initiatives on green car taxes, investment in public transport, and intelligent traffic systems, etc.

Speed restrictions are usually seen as measures of safety for travelling. However, they also offer benefits in terms of energy efficiency. Higher speeds usually mean higher fuel consumption. Hence, adopting maximum speed restrictions force people to drive slower and therefore consume less fuel. In contrary to this, the minimum speed limits on the highways help maintaining a constant traffic flow and therefore reduce the fuel consumption caused by stop-and-go traffic.

3.5.4 Idle Reduction

Idle reduction covers practices and technologies that minimize the amount of time drivers idle their engines. When the engine of a car is not being used to move the car, it can be shut off entirely. Meanwhile, other functions such as accessories and lighting are powered by an electrical source (other than the alternator of the car). Idle reduction offers many benefits, such as; fuel savings, reduced maintenance costs, reduced emissions, and extended life for the car. Applications of such policies may differ from one country or region to another one. For example, in the United States, the Energy Improvement and Extension Act provides tax exemptions for idle reduction technologies. Additionally, SmartWay Transport, a voluntary partnership between the Environment Protection Agency (EPA) and various freight industry sectors, provides education and guidance for truck/locomotive operators, and limited incentives for truck and locomotive idling reduction demonstration projects. Similarly, Canada has the Commercial Transportation Energy Efficiency Rebate Program which is designed to encourage the on-road transportation industry to use equipment for reducing engine idling and cut greenhouse gases [3].

3.6 Exercises

1. What are the primary goals of fuel efficiency standards?
2. What are the types of non-financial incentives applied in transportation sector?
3. Which type of details are provided on fuel-consumption labels?
4. What are AFLEET tool and ATRAVEL tool? For which purposes are they used for?
5. What are the principles that have a clear positive impact on transport energy savings?
6. What is plate restriction scheme? How is it applied?

7. What is the relation between bicycle highways and energy efficiency?
8. What is bus rapid transit system? How is it applied?
9. What is the effect of speed restrictions on energy consumption?
10. What are the types of benefits offered by idle reduction? Explain how idle reduction results in those benefits.

References

1. IEA. (2024). *World energy statistics and balances.* International Energy Agency. Retrieved April 2024, from IEA World Energy Balances https://www.iea.org/data-and-statistics/data-product/world-energy-statistics-and-balances
2. Rodrigue, J.-P. (2024). *4.1—Transportation and energy.* The Geography of Transport Systems. Retrieved April 2024, from https://transportgeography.org/contents/chapter4/transportation-and-energy/
3. IEA. (2024). *Policies database.* IEA. Retrieved April 2024, from https://www.iea.org/policies
4. Doris, E., Cochran, J., & Vorum, M. (2009). *Energy efficiency policy in the United States: Overview of trends at different levels of government.* National Renewable Energy Laboratory. Retrieved April 2024, from https://www.nrel.gov/docs/fy10osti/46532.pdf
5. Marchán, E., & Viscidi, L. (2015). *Green transportation the outlook for electric vehicles in Latin America.* The dialogue. Retrieved April 2024, from https://www.thedialogue.org/wp-content/uploads/2015/10/Green-Transportation-The-Outlook-for-Electric-Vehicles-in-Latin-America.pdf
6. Liu, Z., Berg, M., & Bustad, T. (2020). Review of the existing energy labelling systems and a proposal for rail vehicles. *The Proceedings of the Institution of Mechanical Engineers, Part F: Journal of Rail and Rapid Transit,* 1–11.
7. Fueleconomy.gov. (2024). *Gasoline vehicles: Learn more about the label.* Fueleconomy.Gov. Retrieved April 2024, from https://www.fueleconomy.gov/feg/label/learn-more-gasoline-label.shtml
8. Diesendorf, M., Lamb, D., Mathews, J., & Pearman, G. (2008). *A roadmap for alternative fuels in Australia: Ending our dependence on oil.* The Jamison Group. Retrieved April 2024, from https://www.researchgate.net/figure/Sample-Fuel-Consumption-Label-for-illustrative-purposes-only_fig10_251371507
9. GFEI. (2024). *Vehicle labeling for improved fuel economy.* Global Fuel Economy Initiative. Retrieved April 2024, from https://www.globalfueleconomy.org/transport/gfei/autotool/approaches/information/labeling.asp
10. US DoE. (2024). *Alternative fuels data center.* U.S. Department of Energy's Vehicle Technologies Office. Retrieved April 2024, from https://afdc.energy.gov/tools
11. US DoE. (2024). *Technical assistance.* Office of Energy Efficiency & Renewable Energy. Retrieved April 2024, from https://www.energy.gov/eere/technical-assistance
12. Vehicle Technologies Office. (2024). *Technical assistance. Clean cities and communities.* Retrieved April 2024, from https://cleancities.energy.gov/technical-assistance/
13. EU. (2016). *Action document for institutional capacity building for the transport sector in Uganda.* European Union. Retrieved April 2024, from https://www.eeas.europa.eu/sites/default/files/4._institutional_capacity_building_for_the_transport_sector_in_uganda_-_action_document.pdf

14. Inturri, G., & Ignaccolo, M. (2016). *Making the connection—Energy, transport and urban planning*. Special, Town and Country Planning Association. Retrieved April 2024, from https://www.codema.ie/images/uploads/docs/TCPA_SPECIAL_ExpertP_3.pdf
15. Böhler-Baedeker, S., & Hüging, H. (2012). *Urban transport and energy efficiency*. Deutsche Gesellschaft Für Internationale Zusammenarbeit (GIZ) GmbH
16. Buczynski, A. (2022). *Smart people do not pay twice: Almost half of Flemish cycle highway network overlaps with TEN-T infrastructure*. European Cyclists Federation (ECF). Retrieved April 2022, from https://ecf.com/news-and-events/news/smart-people-do-not-pay-twice-almost-half-flemish-cycle-highway-network
17. National Bus Rapid Transit Institute (NBRTI). (2024). *Bus rapid transit: Elements, performance, benefits*. U.S. Department of Transportation Federal Transit Administration. Retrieved April 2024, from https://www.transit.dot.gov/sites/fta.dot.gov/files/BRTBrochure.pdf
18. Tsivanidis, N. (2018). *The equitable benefits of Colombia's bus rapid transit system*. Voxdev. Retrieved April 2024, from https://voxdev.org/topic/infrastructure/equitable-benefits-colombias-bus-rapid-transit-system

4 Energy Efficiency Policies in Industrial Sector

According to IEA [1], as the end of 2021, the energy consumption in industrial sector constitutes around 30% of total final consumption and sector related emissions constitute more than 18% of CO_2 emissions, globally. Hence, the amount of energy consumed in industrial sector has a huge effect on total demand. The industrial sector uses electricity for operating industrial machinery/motors, computers, lights, etc. It also uses equipment for facility heating, cooling, and ventilation. Hence, it consumes both electrical and heating and cooling related energy. Policies to improve energy efficiency in industrial sector include [2, 3]: regulations and standards, incentives, agreements, reporting and benchmarking, and technical assistance.

4.1 Agreements

Agreements are used widely in the industrial sector in order to meet specific energy use or energy efficiency targets [3]. Either an individual company or an industrial subsector can enter into such agreements, which are mostly voluntary. For example, a voluntary agreement was signed in 1996 with the Luxembourg Federation of Industry in order to increase industrial energy efficiency by 10% during the period 1990–2000. This policy has been prorogated several times with updated targets. Voluntary agreements have also been signed in France with six partners: aluminum, cement industry, fat lime and magnesia lime manufacturers, the French steel federation, glass packaging industry, and a large mail order company. According to these agreements, the companies will need to improve the energy efficiency of their production sites and reduce related emissions [4].

N. Tekbıyık Ersoy, *Energy Efficiency and Renewable Energy Policies*, Synthesis Lectures on Renewable Energy Technologies, https://doi.org/10.1007/978-3-031-64305-7_4

4.2 Incentives

The incentives applied in industrial sector in order to improve energy efficiency can be of two types; financial and non-financial. Financial incentives help industries defray the upfront costs of adopting energy-efficient technologies, while non-financial incentives can offer side benefits such as expedited permitting and job creation. The most straightforward financial instruments are direct incentives, such as; subsidies, tax exemptions and tax rebates [5]. Subsidies can take many forms. They can be direct subsidies, provided to individuals or industry to lower the price of a certain technology, or may take the form of support for extensive research and development (R&D) programs to promote research into innovative energy efficiency solutions. Some other policy examples can be seen in Japan and Netherlands. Green investment tax reduction system in Japan allows the business operators who purchase the target equipment during a specified period to be eligible for special depreciation of 30% against standard purchase prices or 7% tax deduction (only small and medium enterprises). Similarly, in Netherlands, thanks to the tax scheme known as Energy-saving investment credit, the companies can deduct from the taxable profit 41.5% of investments in equipment related to energy conservation and renewable energy [4].

Although direct subsidies may be effective in increasing energy efficiency, they come at the taxpayers' expense. Because the government budget, which is used to provide those subsidies, is mainly formed by citizen taxes. At that point, the appropriate price reduction that is to be provided by the government or related authority should be thoroughly considered. In case of not preferring financial incentives due to economic concerns, non-financial incentives can also be applied. Non-financial incentives can be in the form of energy saving tips, awareness campaigns, energy saving audits, and energy efficiency standards.

4.3 Regulations and Standards

Regulations and standards are mandatory policies designed to improve energy efficiency. They are typically applied to particular types of equipment such as; motors, boilers, electric arc furnaces, rotary kilns, etc. Some countries prefer to adopt minimum energy labelling schemes (MELS) in addition to the standards. MELS usually helps the consumers to compare the energy efficiency of appliances to make more informed purchasing decisions, and Minimum Energy Performance Standards (MEPS) raise the average energy efficiency of these appliances by removing inefficient ones from being sold in the market [6]. For example, MEPS on Motors in Brazil, set the minimum standard for electric motors (three-phase induction motors) as Level IR3. MEPS for electric motors in Chinese Taipei, on the other hand, covers both single-phase induction motors and three-phase induction motors. However, for some types of equipment that have different operating ranges (depending on the process/operation), it can be difficult to set MEPS.

Some regulations can require the industrial facilities to; conduct energy audits, employ an energy manager, and have an energy management system. For example, the Energy Efficiency Act of Austria requires large companies to conduct mandatory energy audits (to be carried at least every four years). Similarly, the Korean government has mandated energy-intensive companies to undertake energy audit on a regular basis. Every five years, companies using over 2,000 toe are obliged to discover energy savings potential and take actions to raise energy efficiency [4].

4.4 Reporting and Benchmarking

Some countries implement programs and policies that promote or require reporting and benchmarking energy consumption. Reporting facility energy use helps in raising management awareness about internal energy consumption trends. This way, the management can decide to take necessary actions in order to reduce the energy consumption. For example, Netherlands has a mandatory system for companies to report energy efficiency actions taken (obligation taken up in the Activities Decree under the Environmental Protection Act). Within the scope of that system, sector specific lists of recognized energy efficiency measures are to be used in reporting by companies. Similarly, in Japan, a new obligation of periodical reporting on the actual status of energy use was established for designated energy-management factories [4].

Benchmarking energy use, on the other hand, provides a means to compare the energy use of a company to that of other companies producing the same products [3]. For example, the Dutch government has the Energy Efficiency Benchmarking Covenant with the industry. In Covenant, energy-intensive industry pledged to be among the world leaders in terms of energy efficiency for processing installations by no later than 2012. As an alternative to that, Energy Efficiency Leader Scheme in People's Republic of China aims to incentivize leaders in energy efficiency (the Top Runners). In order to be a part of this program, industry stakeholders and public institutions must exceed specific energy-efficiency benchmarks set by the China Energy Label [4].

4.5 Technical Assistance

Technical assistance programs are designed to help industries in identifying strategies to reduce energy consumption, such as through energy audits and information campaigns. An energy audit usually includes the assessment of the industrial processes, site visits, measurements and analyses. According to [7], in performing an industrial energy audit, the first step is the initial walk-through, in which the energy audit team or individual auditor, becomes familiar with the facility that they are surveying. The next step is to gather the energy bills and all other current and historical energy related data/information. Examples of such data include [7]:

- General information about the facility (ownership, construction year, product types, scheduled shutdowns, operation schedule, etc.)
- Engineering and architectural plans of the facility
- Energy bills and monthly production for the last 2–3 years
- Climatic data while the auditing is conducted
- Possible archived data with measurements
- Energy management status
- Any energy-saving measures that were implemented

For more information about the details of industrial audits, the interested reader is referred to [8]. Many countries support industrial energy audits in order to reduce the energy consumption. For example, the energy audit support scheme in Sweden is directed towards relatively energy-intensive companies, and provides a support covering 50% of the cost of the energy audit, up to maximum of SEK 30,000. In Korea, Energy Audit Assistance program supports 70% of audit cost for small to medium enterprises (SMEs). The Ministry of Energy, Trade and Industry in Japan encourages energy conservation by providing free energy audit service for small and medium sized factories and business establishments. The energy audit program of Finland consists of promotion of audit activities, development and monitoring, the training and qualification of energy auditors, etc. Business Energy Advice Program (BEAP) in Australia provides a free energy advisory service for eligible small businesses. The service delivers face-to-face, phone and digital advice to small businesses across Australia. This, in turn, helps small businesses realize their energy saving opportunities, choose the best energy plan for their business, and receive advice on energy efficiency opportunities suitable for their industry [4]. Such programs incorporate energy efficiency into systematic decision-making by enabling industries to learn by doing [2].

4.6 Exercises

1. What is the relation between agreements and energy efficiency targets? How does one help another?
2. What are the drawbacks of offering direct subsidies for supporting energy efficiency in industrial sector?
3. What is the major difference between MELS and MEPS?
4. What are the types of financial incentives applied in industrial sector?
5. What are the types of non-financial incentives applied in industrial sector?

6. What are the possible benefits offered by reporting and benchmarking?
7. Which type of data are required for industrial audits?
8. What are the benefits or outcomes of free energy advisory services in industrial sector?

References

1. IEA. (2024). *World energy statistics and balances*. International Energy Agency. Retrieved April 2024, from IEA World Energy Balances https://www.iea.org/data-and-statistics/data-product/world-energy-statistics-and-balances
2. Doris, E., Cochran, J., & Vorum, M. (2009). *Energy efficiency policy in the United States: Overview of trends at different levels of government*. National Renewable Energy Laboratory. Retrieved April 2024, from https://www.nrel.gov/docs/fy10osti/46532.pdf
3. Price, L., & Worrell, E. (2020). *International industrial sector energy efficiency policies*. U.S. Department of Energy Office of Scientific and Technical Information (OSTI). Retrieved April 2024, from https://www.osti.gov/servlets/purl/810469
4. IEA. (2024). *Policies database*. IEA. Retrieved April 2024, from https://www.iea.org/policies
5. Sarker, T., Taghizadeh-Hesary, F., Mortha, A., & Saha, A. (2020). *The role of fiscal incentives in promoting energy efficiency in the industrial sector: Case studies from Asia*. Asian Development Bank Institute. Retrieved April 2024, from https://www.adb.org/sites/default/files/publication/634696/adbi-wp1172.pdf
6. NCCS. (2024). *Responses to feedback and suggestions on Singapore's long-term low emissions development strategy*. National Climate Change Secretariat (NCCS). Retrieved April 2024, from https://www.nccs.gov.sg/files/docs/default-source/default-document-library/annex-for-singapore%27s-leds-public-consultation-response-(final).pdf
7. Peterson, N. (2024). *Industrial energy audit checklist*. LED Lighting Supply. Retrieved April 2024, from https://www.ledlightingsupply.com/blog/industrial-energy-audit-checklist
8. Hasanbeigi, A., & Price, L. (2010). *Industrial energy audit guidebook: Guidelines for conducting an energy audit in industrial facilities*. China Energy Group Energy Analysis Department Environmental Energy Technologies Division. Ernest Orlando Lawrence Berkeley National Laboratory. Retrieved April 2024, from https://electrical-engineering-portal.com/res/Industrial-Energy-Audit-Guidebook.pdf

5 Energy Efficiency Policies in Power Sector

Power sector related energy efficiency policies have two main concerns: production (plant efficiency) and consumption (end-use efficiency). As the utilities have direct contact with the consumers and knowledge of the consumers' needs in relation to electricity production and distribution, they can utilize this knowledge to develop pricing policies and incentive programs. Such programs can help maintain the balance between demand and supply on a temporal basis (hourly, daily, etc.) [1]. The types of energy efficiency policies in power sector include: Regulations and Plans, Codes and Standards, Incentives and Technical Assistance.

5.1 Codes and Standards

Codes and standards in power sector are similar to those explained for other sectors. They can be in the form of minimum energy performance standards for transformers (e.g. in Vietnam), or energy efficiency portfolio standards (EEPS) (e.g. in Illinois), which are also called energy efficiency resource standards (EERS). EEPS/EERS is a performance-based standard for utility energy efficiency programs. It is mainly a policy requiring electricity or natural gas utilities to achieve specified levels of customer energy savings. According to [2], the typical EERS objectives are percentage savings of electricity or natural gas sales, as compared to a business-as-usual case. In some cases, EERS may also have goals towards reducing peak electricity demand. Some EERS impose electric utilities to procure a specified amount of energy efficiency or demand side management resources for managing and reducing energy usage and demand by consumers. Many utilities around the world started applying demand side management (DSM) programs to help save energy. DSM

N. Tekbıyık Ersoy, *Energy Efficiency and Renewable Energy Policies*, Synthesis Lectures on Renewable Energy Technologies, https://doi.org/10.1007/978-3-031-64305-7_5

covers both demand response and long-term or permanent energy efficiency measures. DSM strategies include; energy efficiency/energy conservation, peak demand clipping, demand valley filling, load shifting, flexible load shaping, and strategic load growth. More detailed information about these strategies can be found in [3].

5.2 Direct Financial Incentives

One of the key methods for reducing energy consumption is to provide incentives for the end-users and utilities. Hence, in addition to the measures mentioned above, some governments or utility companies offer rebates, grants and loans to help subsidize the cost of purchasing energy efficient technologies. For example, in Hungary, the National Energy Conservation Programme offers energy efficiency grants to households. Different types of energy efficiency improvements are subsidized as part of the programme, such as; change or insulation of windows and doors, improvement of heating and hot water supply, thermal insulation of existing buildings, etc. Subsidies change based on the type of energy efficiency improvements (15–20%) [4].

In some cases, on-bill financing programs can also be applied. On-bill financing refers to a loan made to a utility customer to pay for energy efficiency improvements to the customer's house or building. With such programs, the utilities help the customers to invest in energy efficiency improvements, such as; installing new lighting, upgrading to a high-efficiency air conditioner or adding insulation, etc. An on-bill program may be administered by the utility directly or by an outside administrator in conjunction with the utility. The process works as follows: Until the loan is repaid, monthly payments are collected by the utility on the bill. Utilities require consumers to either repay the costs through a loan (which they would need to repay even if they moved), or through a higher tariff (which would remain with the house). On-bill loans broaden customer eligibility. Some customers willing to invest in energy efficiency improvements may not actually be eligible for a conventional loan or may simply find the loan too expensive. However, as the on-bill loan has bill neutrality (the customer's average monthly payments are not expected to increase), this can motivate the customer to invest in efficiency improvements. Although billing a loan payment in connection with the utility bill can increase customers' convenience, it may also have undesired results for the customers. Many on-bill programs allow the utility company to suspend service to the customers who fail to make their loan payments. More details about on-bill loans, such as challenges and questions to consider can be found in [5].

The end-users are already motivated to reduce their energy consumption, as this would reduce their electricity bill. Utilities, on the other hand, do not have such motivation as the utility revenues are based on volumetric sales of electricity. This creates a financial incentive to sell greater volumes of electricity (known as throughput incentive), and therefore, can be considered as a disincentive for the utilities to support energy-efficiency programs. The solution to this problem can be disassociating profits from the sales volume, and linking them to the reductions in energy demand. One successful regulatory measure is to apply decoupling. Under decoupling, the utilities receive differing rates per kWh depending on the total electricity demand [1]. For example, if the electricity demand increases above a pre-identified target, the rates fall. Contrarily, if the demand decreases, the rates rise. Hence, in order to receive higher per kWh rates, the utilities need to help the end-users to reduce their energy demand. This may include providing end-user incentives (such as subsidized energy audits) or by providing technical assistance and information on ways of reducing the demand.

5.3 Demand Based Incentives

Some examples of the end-user related incentives can be listed as follows [1]; direct financial incentives for energy efficient technologies and making the price of electricity dependent on the time of day and/or the total electricity usage. As direct financial incentives have been explained in previous section, this section will discuss the latter. The first aim is to reduce the peak demand. In order to be able to apply the incentives for reducing the peak demand, the users should have either interval meters or smart meters. There are mainly three types of meters; accumulation meters, interval meters and smart meters. Accumulation meter is the oldest and the most common type of meter. An accumulation meter records the total amount of energy used over a period of time (about a predetermined number of months). In order to determine the energy consumption, a meter reader must go to the customer's property and physically read the meter. An interval meter, on the other hand, records energy use over short intervals, typically every 30 min. The advantage of an interval meter over accumulation meter is the ability of measuring when as well as how much energy the customer uses. This allows the retailers to offer prices and deals based on the time the customer uses power. However, a meter reader must still go to the customer's property to physically read the meter. The best option among the meters is the smart meter, as it can record when and how much electricity is used and allows wireless communication between the electricity supplier and the meter, eliminating the need for a meter reader. This way, one can even read the information recorded by their smart meter, by using an online portal or an in-home electronic display. This information helps the user to understand when and how much electricity he/she uses, and identify the ways to save electricity. Getting more frequent and real time data also helps the utilities to offer complex pricing structures that can be useful in manipulating the demand, and apply demand response programs. Demand response is about altering the energy demand in response to available energy supply by financially incentivizing users to make short-term reductions in energy demand. More details about demand response programs can be found in [3].

Making the price of electricity dependent on the time of day helps in reducing the peak demand. One of the methods to do that is to apply Time of Use (TOU) pricing. Based on

TOU, when utility customers consume energy at different time intervals of the day, they pay different electricity prices. TOU prices may change from one season to another, or one region to another. A real life example of that change can be seen in [6]. Figure 5.1 illustrates an example of TOU. The morning and late afternoon periods are much more costly in terms of electricity usage. This may be due to the following reasons: many people wake up around the same time and start using electricity, and this creates a peak demand in the morning. Similarly, in the late afternoon, many people return from work or school and they create another peak period. In order to reduce that demand, the price is increased. In contrary to these, in the night time, many people sleep and therefore the energy demand is low. This allows the utility companies to reduce the price of electricity in order to motivate people to use their devices on those periods rather than the peak periods.

Although TOU pricing can be very successful in terms of shifting the demand to other times of the day, it does not reduce the total energy consumption of the consumer. It just allows to reduce the peak demand and the utility bill paid by the consumer. In some cases, reducing the total energy demand can be a necessity; for example; in some cases where the current power generation capacity is not enough for the current power demand, or if this will happen in the near future. In such cases, tiered-pricing structure is preferred. A tiered-pricing structure increases the cost of electricity for incrementally larger blocks of electricity consumption. Imagine a case in which the price of electricity for the first 300 kWh consumed in a month is \$0.07/kWh, whereas the price of electricity consumed for the next 300 kWh in that month is \$0.11/kWh. In such a case, most of the end-users

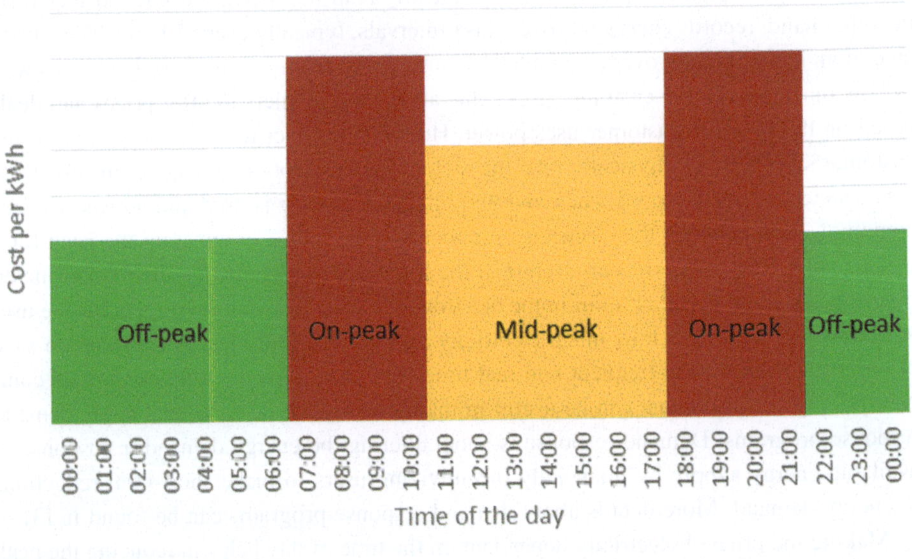

Fig. 5.1 An example TOU schedule for winter

would naturally try to consume less than 300 kWh (if possible), as otherwise they would be paying more per kWh.

5.4 Regulations and Strategic Plans

Regulations applied in a country are usually tailored to that country's energy needs. These regulations can be specific to power sector or may be general regulations covering some goals about the power sector (in addition to the other sectors). For example, the energy efficiency law in Turkiye requires certain energy users to report on their energy consumption to the General Directorate of Renewable Energy (GDRE). These obligations apply to many sectors (Industrial establishments, Commercial or service buildings, power sector, etc.). The one in power sector covers power plants which have minimum 100 MW installed capacity [4]. Even those regulations specific to power sector can focus on more specific issues in power sector, such as smart grids. As the awareness about the smart grids and smart meters increase, the related regulations also increase in amount. Many countries, such as Brazil, Austria, Canada have specific regulations about smart meters. The regulation in Brazil mandates local electricity distribution companies to install smart meters for their customers. The one in Austria sets the technical and functional requirements for smart meters installation. Some regulations also set descriptions of smart grid technologies to be implemented, such as the one in Canada.

Strategic plans include policy signals that demonstrate national plans for reaching energy efficiency, including national targets and creation of institutions. It should be remembered that setting a target does not necessarily guarantee that it can be reached. In order to reach the specified target, a country should have a strategic plan determining what will be done, how it will be done and when. The plan should also have intermediate targets to be reached, and possible methods to be used in reaching those targets. This would allow the transition to happen in a gradual way. For example, the European Commission proposed the REPowerEU Plan which aims at decreasing the European Union's dependency on Russian fossil fuels [4]. Although most of the measures listed in the plan are about renewable energy, there are also some measures about energy efficiency, such as increasing the EU's 2030 binding energy savings target to 13%. The main advantages of targets are that they are clear and measurable. Moreover, they can be used for long-term planning of other policies for meeting these targets.

5.5 Technical Assistance

Technical assistance programs in power sector are very similar to those explained before for other sectors. They are mainly based on providing technical information to those who need it. Technical assistance is particularly needed when the power sector and the environmental ministries lack a deep understanding of the dynamics of energy efficiency policy and the main drivers of energy demand. Especially in situations in which the utilities are unaccustomed to promoting energy efficiency, technical assistance can help in either of the following steps [7]:

1. Evaluating the savings potential and resulting avoided energy costs
2. Evaluating specific energy efficiency measures and cost-effectiveness of implementation
3. Developing a plan for implementation
4. Measuring the impacts through robust evaluation protocols

Technical assistance may include activities to support development of a robust energy service company (ESCO) sector, including monitoring and verification training, handbooks, and competency standards. It may also cover workshops, training, and online tools to facilitate energy management systems. There are successful technical assistance programs in the United States. For example, the Rural Energy for America Program (REAP) applied in the United States, includes the Energy Audit and Renewable Energy Development Assistance Grant Program. Moreover, the U.S. Environmental Protection Agency (EPA) provides technical assistance to state utility regulators who want to explore utilization of clean energy. This technical assistance allows the states to learn from each other and pursue best practice policies and programs for energy efficiency, and clean distributed generation [4].

5.6 Exercises

1. What are energy efficiency resource standards? For which purposes are they used for?
2. What are the types of demand side management strategies?
3. What is on-bill financing? What are the possible benefits that it offers to the consumers?
4. What is the relation between throughput incentive and decoupling? How does one affect another?
5. What is the main difference between accumulation meters and smart meters?
6. How do smart meters help in developing policies that can motivate reducing peak demand?
7. What are the major criteria used to determine the prices in TOU pricing?

8. What is the benefit of adopting a tiered-pricing structure in terms of energy consumption?
9. What is the effect of strategic plans on energy efficiency targets?
10. What are the types of activities included in technical assistance for power sector?

References

1. Doris, E., Cochran, J., & Vorum, M. (2009). *Energy efficiency policy in the United States: Overview of trends at different levels of government*. National Renewable Energy Laboratory. Retrieved April 2024, from https://www.nrel.gov/docs/fy10osti/46532.pdf
2. ASE. (2024). *Energy efficiency resource standard (EERS)*. Alliance to Save Energy. Retrieved April 2024, from https://www.ase.org/sites/ase.org/files/resources/Media%20browser/eers_fact_sheet_9-13.pdf
3. Tekbıyık-Ersoy, N. (2023). Demand-side management and demand response. In A. Moreno-Munoz & N. Giacomini (Eds.), *Energy smart appliances: Applications, methodologies, and challenges* (pp. 93–116). IEEE.
4. IEA. (2024). *Policies database*. IEA. Retrieved April 2024, from https://www.iea.org/policies
5. Henderson, P. (2013). *On-bill financing: Overview and key considerations for program design*. Natural Resources Defense Council (NRDC). Retrieved April 2024, from https://www.nrdc.org/sites/default/files/on-bill-financing-IB.pdf
6. Oshawa Power. (2024). *Time of use*. Oshawa Power. Retrieved April 2024, from https://www.oshawapower.ca/time-of-use/
7. Mcneil, M., Can, S. D. L. R. D., & Gonzalex, A. D. (2020). *Scaling up energy efficiency in developing countries: The building blocks of energy efficiency*. U.S. Agency for International Development. Retrieved April 2024, from https://ee4d.org/wp-content/uploads/sites/40/2021/05/USAID_EE4D_Energy-Efficiency_Building-Blocks_Toolkit_508.pdf

6 Comparative Analysis of Energy Efficiency Policies

6.1 Comparative Analysis Based on Sectors

This section provides a comparative analysis of global energy efficiency policies, conducted for this book. The analysis is conducted separately for each sector; building, transportation, industrial, and power. The main goal of performing such an analysis is to understand which regions and which countries have more energy efficiency policies for supporting energy efficiency in those sectors. This can be helpful in understanding which sectors the countries are paying more attention to, and why. One may argue that the number of policies do not necessarily indicate the level of devotion to energy efficiency, as a country can have one policy covering many different issues. However, repetitive action indicates dedication. Also, having many energy efficiency policies in a sector, means that the country is offering a diverse set of support mechanisms. Diversity is the key to peak productivity, as one policy may not be appealing to one consumer/producer and another type of policy can.

This analysis is performed by using International Energy Agency's (IEA) energy policies database [1]. In that database, the following filters have been applied: Topic: Energy Efficiency, Status: In force. After that, in order to determine the policy leaders for each sector, the filters shown in Table 6.1 are applied. It should be noted that Economy-wide (multi-sector) criteria is not included in this analysis. After applying the above-mentioned criteria, the top three regions have been determined in terms of number of energy efficiency policies. As seen from the table, in all the sectors, Europe is the leader in terms of number of EE policies. This can be due to EU energy efficiency directive and EU's 2030 targets requiring 32.5% energy efficiency improvement until 2030. The top three regions hardly change form one sector to another. It is also interesting to see that United States and Canada are among the top three countries in terms of adopted EE policies, irrespective of the sectors. This indicates their dedication for supporting EE with as many policies as

N. Tekbıyık Ersoy, *Energy Efficiency and Renewable Energy Policies*, Synthesis Lectures on Renewable Energy Technologies, https://doi.org/10.1007/978-3-031-64305-7_6

Table. 6.1 Energy efficiency policy filtering criteria and the regional and country based leaders

	Building	Transportation	Industrial	Power
Sector	• Buildings • Residential • Services • Existing buildings and retrofits	• Transport • Road transport • Passenger transport (road)	• Industrial	• Power, heat and utilities
Top 3 regions	1. Europe 2. Asia Pacific 3. North America	1. Europe 2. Asia Pacific 3. North America	1. Europe 2. Asia Pacific 3. North America	1. Europe 2. North America 3. Asia Pacific
Top 3 countries	1. United States 2. Canada 3. India	1. France/United Kingdom 2. Canada 3. Germany/United States	1. United States 2. Germany 3. Canada	1. United States 2. Canada 3. Poland

possible. As it can be seen in transportation sector, the leader position is shared between France and United Kingdom as they have the same amount of policies. The same is valid for Germany and United States, as they share the same number of policies. United States seems to be the world leader in terms of adopted EE policies in almost all of the sectors, except the transportation sector.

The next four figures show the breakdown of the top three regions supporting EE in the above-mentioned sectors. In order for the reader to easily detect the analysis results of the sector that they require, the figures are constructed by using different color scales. Figure 6.1 shows the breakdown of the top three regions supporting EE in building sector. It reveals the top 10 countries in each of those regions. As seen from the figure, Europe has a more competitive landscape as the number of EE policies are close to each other for the top 10 countries. This shows that many countries in EU region are well motivated in supporting energy efficiency. In Asia Pacific, however, majority of the policies are adopted in India, Australia and China. In North America region, United States is by far the leader. But it should be noted that Canada itself, which is the second country in North America region, has more policies than the leaders of Europe and Asia Pacific.

Figure 6.2 shows the top 10 countries in each of the leader regions supporting EE in transportation sector. As seen from the figure, once again, Europe shows its diversity. Many European countries are supporting EE in transport with many different measures. In Asia Pacific, top three changes to be China, Japan and Australia. In North America region, Canada surpasses United States for the first time, which means they pay more attention to the transportation sector when compared to United States.

Observing Figs. 6.3 and 6.4 reveals that Europe's diversity consistency continues in other sectors. Another interesting finding is Indonesia in Asia Pacific region for the industrial sector policies. Considering the highly populated countries in the region, such as India (1.417 billion) and China (1.412 billion) (based on the total population data (2022) provided in [2]), Indonesia (275.5 million) seems to be much more dedicated in terms

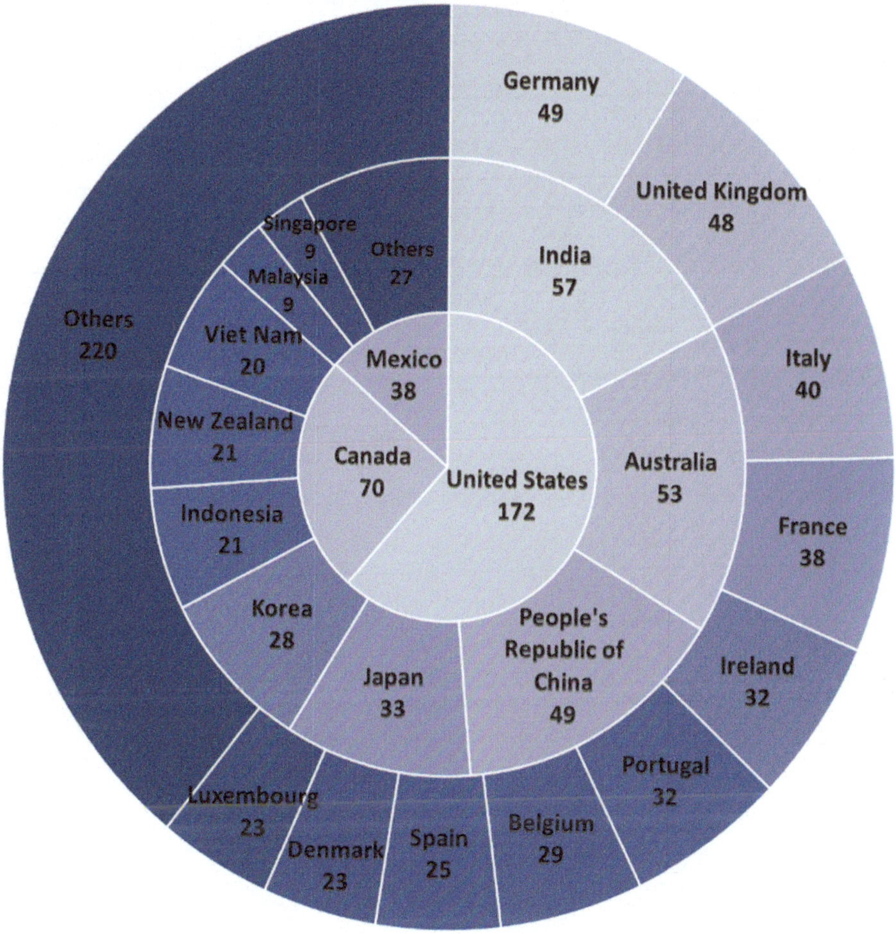

Fig. 6.1 Country based breakdown of the top three regions for EE policies in building sector

of increasing EE in industrial sector. In power sector, the number of policies seems to be much lower than the policies adopted in other sectors. This indicates that much more policies are needed in order to boost EE in power sector.

6.2 Exercises

1. Which regions are among the top three in power sector?
2. Which region is the leader (has most policies) in almost all the sectors?
3. Which countries are among the top three in almost all the sectors?
4. What could be reasons behind Europe's sector-based diverse policy adoption?

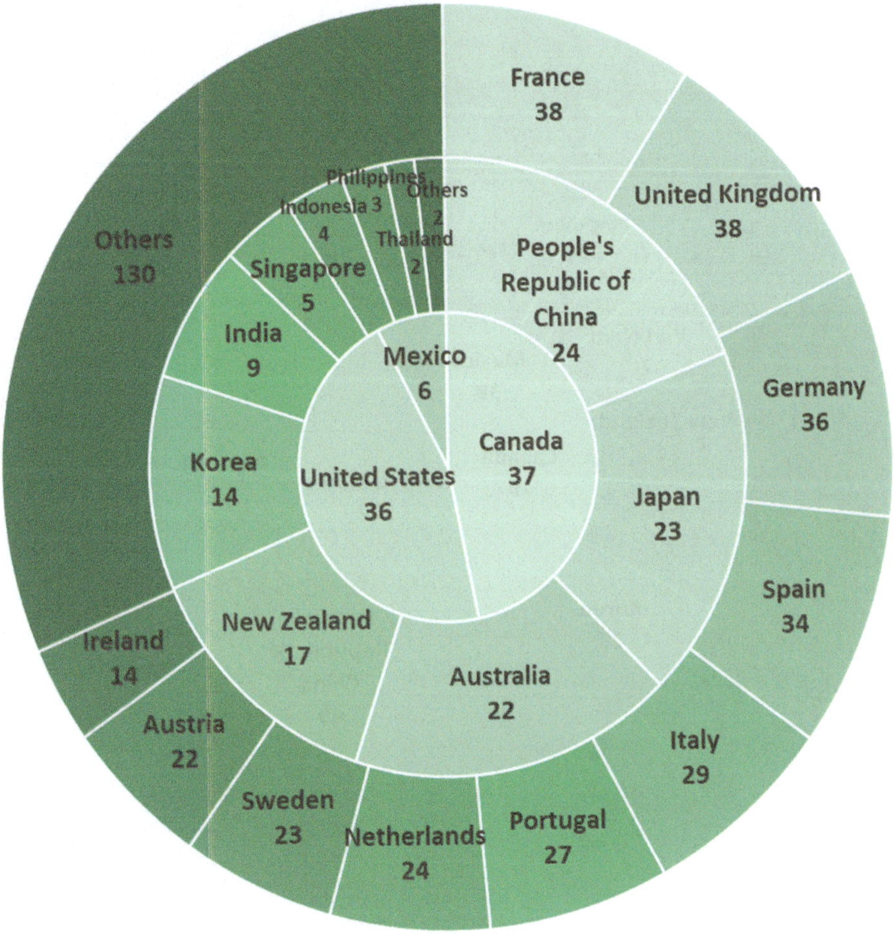

Fig. 6.2 Country based breakdown of the top three regions for EE policies in transportation sector

5. In which sector, the more EE policies are adopted? Why?
6. Which countries are the regional leaders (leader in their region) in terms of number of EE policies adopted in building sector?
7. Which countries are the regional leaders (leader in their region) in terms of number of EE policies adopted in transportation sector?
8. Which countries are the regional leaders (leader in their region) in terms of number of EE policies adopted in industrial sector?

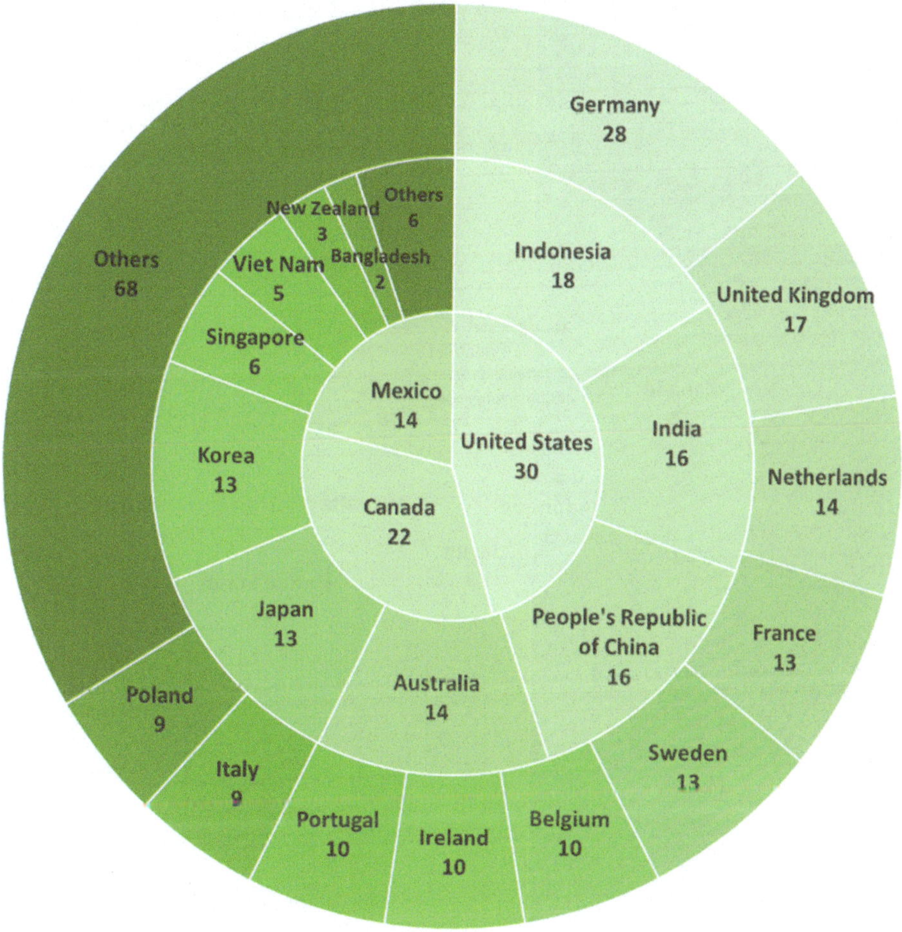

Fig. 6.3 Country based breakdown of the top three regions for EE policies in industrial sector

9. Which countries are the regional leaders (leader in their region) in terms of number of EE policies adopted in power sector?
10. Why having more EE policies is important? What is the effect of having more policies on energy consumption?

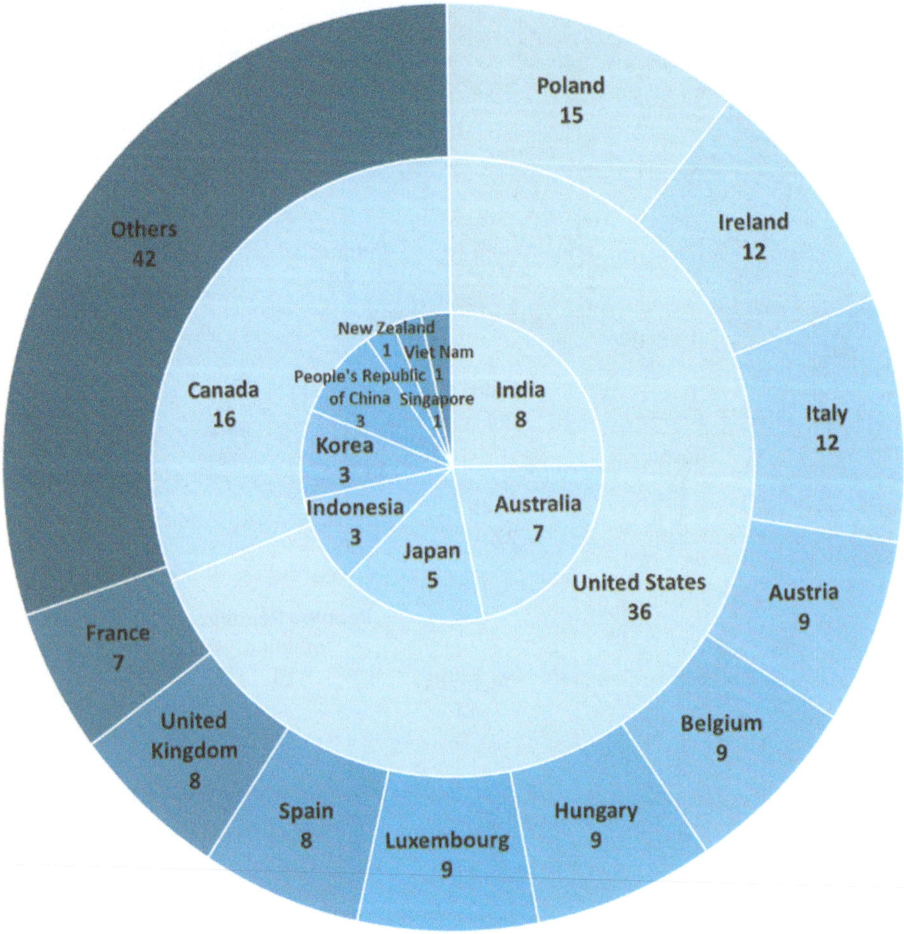

Fig. 6.4 Country based breakdown of the top three regions for EE policies in power sector

References

1. IEA. (2024). *Policies database.* IEA. Retrieved April 2024, from https://www.iea.org/policies
2. The World Bank. (2024). *World development indicators.* Databank. The World Bank. Retrieved April 2024, from https://databank.worldbank.org/reports.aspx?source=world-development-indicators

7 Renewable Energy and the Need for Renewable Energy

7.1 Renewable Energy

Renewable energy (RE) means energy from renewable sources, such as; solar, wind, geothermal, tidal, wave and other ocean energy, hydropower, biomass, landfill gas, sewage treatment plant gas, and biogas [1]. These resources are called renewable as they are naturally replenished in a short period of time. Solar and geothermal energy can be used both for electricity production and for heating and cooling. There are various solar technologies, such as Photovoltaics (PV), passive solar, solar water/air heating, and concentrating solar power. Similarly, geothermal energy can be used in various forms: geothermal power plants, direct use of geothermal heat, and ground-source heat pumps.

Some other renewables, such as wind, tidal, wave and hydropower are usually used for producing mechanical motion or electricity. Types of wind turbines used for electricity production can be listed as horizontal axis wind turbines (HAWTs) and vertical axis wind turbines (VAWTs). Wind energy can also be used for transportation, such as sailing. When it comes to hydropower, many different types of power plants can be used to harvest the power of water. Examples can be listed as; impoundment, diversion, and pumped-storage. Similarly, tidal power plants can be in various forms; tidal barrage, tidal stream generator, and dynamic tidal power (DTP). Wave energy can also be harvested. In order to do that, wind energy converters are needed. Some of the most common wind energy converters can be listed as; point absorbers, attenuators, oscillating wave surge, and oscillating water column.

Biomass can be considered as the most diverse RE source, as it can be used in almost all the sectors; electricity, heating and cooling, and transportation (in form of biofuels). Biomass sources for energy include; Wood and wood processing waste, agricultural crops and waste materials, biogenic materials in municipal solid waste and animal manure and

N. Tekbıyık Ersoy, *Energy Efficiency and Renewable Energy Policies*, Synthesis Lectures on Renewable Energy Technologies, https://doi.org/10.1007/978-3-031-64305-7_7

human sewage for producing biogas. According to [2], biomass is converted to energy through various processes listed below:

- Direct combustion for producing heat
- Biological conversion for producing liquid and gaseous fuels
- Chemical conversion for producing liquid fuels
- Thermochemical conversion for producing solid, gaseous, and liquid fuels

Renewable energy offers many benefits, which can be summarized as follows [3]:

- Reducing carbon emissions from energy production
- Reducing air pollution
- Enhancing reliability and security of the power grid
- Job creation for production and manufacturing of RE technologies
- Decreasing energy imports
- Expanding energy access for remote or isolated communities.

7.2 The Need for Renewable Energy

Multiple independent studies found that between 90–100% of the scientists agree that humans are responsible for climate change (with most of the studies finding a 97% consensus) [4]. Actually, this is not the first climate change happening on Earth. Natural changes in the sun's activity or large volcanic eruptions have caused climate change before, leading to the shifts in Earth's temperatures and changing weather patterns. However, since the industrial revolution (almost in the last 200 years), the main driver of global temperatures was not the natural causes. In industrial revolution (from about 1760 to 1840), something revolutionary happened; a transition from creating goods by hand to using machines. For more detailed information about industrial revolution, the interested reader is referred to [5]. What makes the industrial revolution important is that for most of the human history, people relied on very basic forms of energy, such as; human or animal muscle, and the burning of biomass (such as wood or crops). But the industrial revolution revealed the enormous potential of a whole new energy resource; fossil fuels

Fossil fuels are petroleum (oil), natural gas and coal. They are formed from the decomposition of buried carbon-based organisms that died millions of years ago. That's why they are called "fossil" fuels. Their formation requires heat, pressure and time (millions of years). Through that time, they create carbon-rich deposits that can be extracted and burned for energy. However, due to their carbon-rich nature, burning fossil fuels causes significant amount of pollution and greenhouse gas (GHG) emissions, and those emissions enhance the greenhouse effect in return. The greenhouse effect is the natural warming of the earth that happens when gases in the atmosphere trap heat from the sun that would otherwise escape into space. Hence actually, the natural greenhouse effect process makes the earth habitable. However, this process did not stop operating when more GHGs have been injected into the atmosphere due to fossil fuel combustion. According to [6], for most of the past 800,000 years (much longer than the existence of human civilization), the concentration of CO_2 in Earth's atmosphere was roughly between 200 and 280 parts per million (ppm). However, as of 2023, it has reached more than 420 ppm, which is 50% higher than preindustrial levels. The current higher concentrations of greenhouse gases, and carbon dioxide (CO_2) in particular, are causing extra heat to be trapped and average global temperatures to rise. According to [7], the Intergovernmental Panel on Climate Change (IPCC) has found that emissions from fossil fuels are the dominant cause of global warming. Driven by the burning of fossil fuels, global warming is altering the earth's climate system in many ways [6]:

- Causing more frequent and intense extreme weather events (heat waves, hurricanes, droughts, and floods)
- Making wet regions wetter and dry regions drier by intensifying precipitation extremes
- Raising sea levels (due to melting ice sheets and glaciers)
- Increasing ocean temperatures
- Causing changes in ecosystems and natural habitats; animals' geographic ranges, and seasonal activities, etc.

In addition to the problems stated above, there is also the energy demand–supply problem which is becoming more challenging due to increasing population and people's energy-intensive lifestyles. When energy demand increases, the supply should be increased as well. However, most of the World's energy is still being produced by using fossil fuels, which are non-renewable due to the long time span required for their formation. Figure 7.1 illustrates the total energy supply by source as of 2021. The figure is constructed based on the data provided in [8].

As seen from the figure, the share of fossil fuels in total energy supply is around 80%. Hence, in order to provide environmentally friendly and sustainable energy for the people, a transition toward renewable energy is needed. Considering the urgency of the problems mentioned above and the benefits of RE, investments to RE and RE production should be increased significantly in a short amount of time. Hence, renewable energy support mechanisms and policies are needed in order to boost RE deployment. Next chapter will explain different types of renewable energy policies that can be used for this purpose.

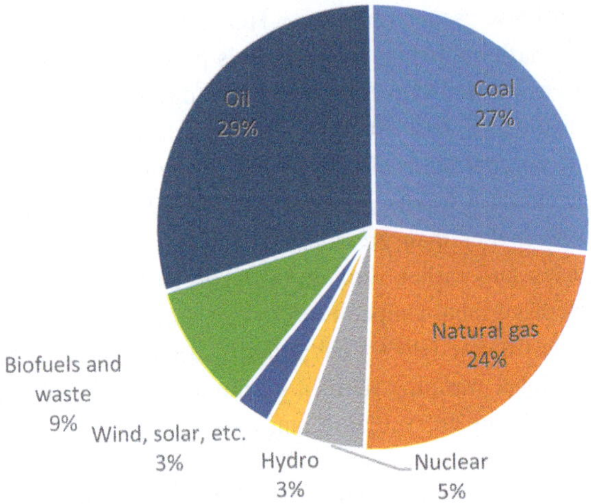

Fig. 7.1 Total energy supply by source as of 2021

7.3 Exercises

1. What are the types of renewable energy?
2. What are different types of solar energy technologies?
3. Which RE source is the most versatile one in terms of area of usage?
4. What are the benefits of renewable energy?
5. What is the main relation between fossil fuels and global warming?
6. What are the effects of global warming on the Earth's climate system?
7. What is the share of renewables in total energy supply?

References

1. European Parliament. (2024). *Renewable energy: Setting ambitious targets for Europe*. European Parliament. Retrieved April 2024, from https://www.europarl.europa.eu/topics/en/article/20171124STO88813/renewable-energy-setting-ambitious-targets-for-europe
2. EIA. (2023). *Biomass explained*. U.S. Energy Information Administration. Retrieved April 20224, from https://www.eia.gov/energyexplained/biomass/
3. Office of Energy Efficiency & Renewable Energy. (2024). *Renewable energy*. Office of Energy Efficiency & Renewable Energy. Retrieved April 2024, from https://www.energy.gov/eere/renewable-energy
4. United Nations. (2024). *The facts on climate and energy*. United Nations. Retrieved April 2024, from https://www.un.org/en/climatechange/science/mythbusters
5. National Geographic. (2024). *Industrialization, labor, and life*. National Geographic Headquarters. Retrieved April 2024, from https://education.nationalgeographic.org/resource/industrialization-labor-and-life/

6. Denchak, M. (2023). *Greenhouse effect 101*. NRDC. Retrieved April 2024, from https://www.nrdc.org/stories/greenhouse-effect-101#solution
7. Clientearth Communications. (2022). *Fossil fuels and climate change: The facts*. Clientearth. Retrieved April 2024, from https://www.clientearth.org/latest/news/fossil-fuels-and-climate-change-the-facts/
8. IEA. (2024). *Energy statistics data browser.* International Energy Agency. Retrieved April 2024, from https://www.iea.org/data-and-statistics/data-tools/energy-statistics-data-browser?country=WORLD&fuel=Energy%20supply&indicator=TESbySource

8 Renewable Energy Policies

Policies that are usually adopted for the development of renewable energy can be categorized into three groups: Incentives, education and research, public finance and regulations. Incentives are those that provide incentives to the customers/consumers, while public finance is more about the efforts of public or private organizations, such as banks, etc. Regulations, on the other hand, are the rules, laws, legislations, etc. designed to boost the renewable energy development, based on some pre-determined rules and regulations set by the government (or the related authority). The diagram provided in Fig. 8.1 (which is constructed mostly based on the data provided in [1]) illustrates the related policies in each category. The next sections are devoted to the explanation of each of these policies.

8.1 Incentives, Education and Research

The types of incentives applied in renewable energy development are explained in this section, along with the policies that support RE education and related research.

8.1.1 Capacity Building

IRENA Secretariat [2] defines capacity building as the process of enhancing individual skills or organizational competence and also nurturing the supportive patterns of social relationships. Capacity building requires individuals acquiring technical/managerial expertise and performance capabilities. It may also require communities and organizations to acquire systems, processes and structures to be more efficient and effective. An interested

N. Tekbıyık Ersoy, *Energy Efficiency and Renewable Energy Policies*, Synthesis Lectures on Renewable Energy Technologies, https://doi.org/10.1007/978-3-031-64305-7_8

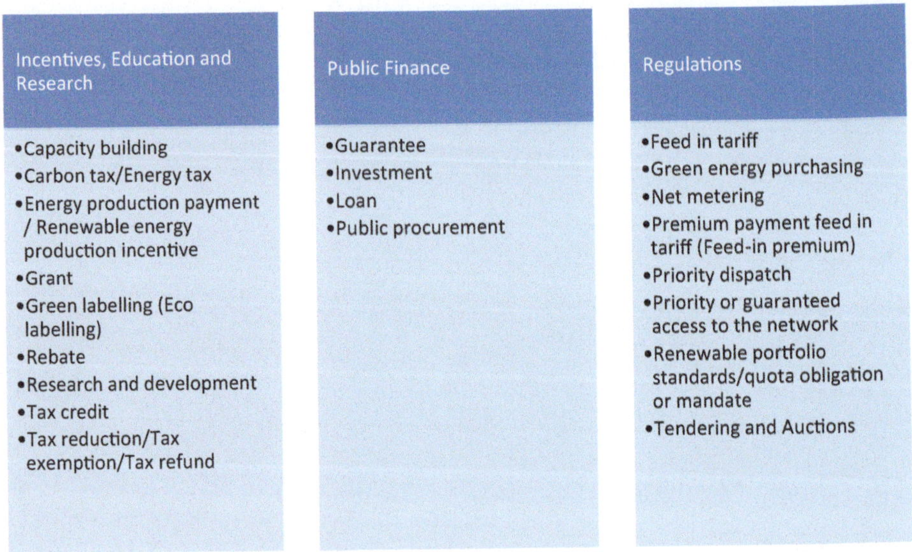

Fig. 8.1 Categorization of renewable energy policies

reader can find more information about capacity building at the individual level, organizational level and system level in [2]. When it comes to sustainability, capacity building involves acquiring the necessary skills and knowledge to be able to implement sustainable and environmentally friendly practices. Mungai [3] claims that sustainability is complex and organizations may find it difficult to navigate this complex environment and to put effective sustainability efforts into place without the required capacity. Capacity building helps individuals and organizations to identify areas in which they can reduce their negative environmental and social impact.

8.1.2 Carbon Tax/Energy Tax

A carbon tax is mainly a pricing mechanism that places a fee or tax on greenhouse gas emissions caused by burning fossil fuels (oil, natural gas, coal). In this scheme, the government sets a price that emitters must pay for each ton of greenhouse gas emissions they emit. Carbon taxes are applied only in some countries while energy taxes are applied in many countries. Energy taxes are an important source of revenue in many countries and, they can be the dominant source of government revenue. The most common form of energy taxes are fuel excise taxes [4]. Fuel excise taxes are similar to carbon taxes as the tax liability increases with the increasing use of the fossil fuels. However, mostly, they only apply to certain fuels, and therefore differ from the carbon taxes. More detailed information about taxing on energy use can be found in [4].

8.1.3 Energy Production Payment (Renewable Energy Production Incentive)

Energy production payment is a direct payment from the government for each unit of renewable energy produced. It is also known as renewable energy production incentive (REPI). For example, the Energy Policy Act of 1992 (EPAct 1992) in the United States offers a production incentive payment of $0.015/kWh for publicly-owned utilities (that cannot benefit from production tax credit (PTC) as they pay no federal income taxes) [5].

8.1.4 Grant

Grant is a financial help from the government that does not have to be repaid. It is usually given for specified purposes to an eligible recipient, and requires certain qualifications related with the use and maintenance of specified standards.

Grants help in reducing the investment costs related with the purchase or construction of renewable energy equipment or infrastructure. Some grants, like the one offered by the Commonwealth Wind Incentive Program [6], may help in paying for site assessments, feasibility studies and design and construction of the sites. While some others, like the Community Batteries for Household Solar grants in Australia, aims to support the deployment of community batteries. There can also be technology specific grants. For example, the European Regional Development Fund (ERDF) Energy Grant Scheme in Malta provide grants for electricity generation from renewable sources such as solar and wind [5].

8.1.5 Green Labelling (Eco Labelling)

Green labelling is usually a government-sponsored labelling that guarantees that energy products meet certain sustainability criteria. The aim is to facilitate voluntary green energy purchasing [1]. Labelling schemes inform consumers about products (such as electricity source) and influence consumer market activity, as a result of enhanced knowledge level of the consumer. Producers who would like to use a label on their products must comply with some standards and usually pay a license fee. According to [7], green labelling schemes require an external auditor to examine whether the certified company complies with the standard. This external auditor has the authority to ask for corrective measures if needed and, if there is non-compliance, to withdraw the certificate. Green labels, also referred as Eco-labels, symbolize beneficial consumer choices in terms of quality, health, environmental, or other matters. This way, green labelling is expected to steer consumers towards environmentally friendly products, and encourage the manufacturers to develop such products. For example, some governments require labelling on consumer bills, with the disclosure of RE share in the energy mix. This allows the consumer to realize the effect of their energy consumption on the environment, and may be to decide lowering down the energy consumption, investing in environmentally friendly technologies or installing solar panels. The reader interested in roles of consumers with regard to green labels is referred to [7].

Increased use of green labels (or eco labels) can reduce the environmental impacts of the manufacturing industries, and advance sustainable consumption and production patterns. However, if this process is considered from the producers' point of view, the costs to conduct an environmental impact analysis or life-cycle analysis increase the production costs. There are also other challenges such as; the lack of expertise, the lack of financial and organizational capacity to conduct these analyses, especially in developing countries. Hence, unfortunately, not all eco-labels have a transparent standard-setting process, a robust verification scheme, or scientifically validated standards. According to [8], the required design and implementation principles for effective eco-labelling, are as follows:

- The consumer must understand and trust the information delivered
- Process of setting standards should be transparent and inclusive
- There should be a mandatory review mechanism set within the scheme, based on the changes in scientific knowledge and the signals from the market
- There should be a robust verification scheme and simple certification process
- The awarding organization should be unbiased and independent
- Participation should be open to all kinds of organizations

A general classification of labelling is provided in [9, 10] as; Eco labelling Type I, Eco labelling Type II and Eco labelling Type III. According to [9], Type I eco-labels are voluntary, multi-criteria certification programs that award labels to environmentally friendly products (based on life cycle considerations), when compared to similar products within the same category. Such labelling programs have been introduced in several countries to help consumers identify environmentally friendly products. Some examples of type I eco-labels are listed in Table 8.1, along with the countries in which they are used and the related awarding criteria. In Fig. 8.2, Nordic Swan Ecolabel can also be observed.

Type II Eco Labels (also called Environmental Claims) are symbols or statements that describe environmental characteristics of a product. They are self-declarations made by the manufacturers themselves, and are commonly used to indicate less environmental impact compared to a standard product. For example, in case of electricity, the supplier aims to convince the consumer that the product has a lower environmental impact, by adding statements like; "Green electricity", "Ecoelectricity" or "100% hydropower". In these cases, the consumer is expected to trust the supplier that the self-defined environmental claim is correct [10]. Although such eco-labels are not certifiable by third parties, the information they provide must be accurate and verifiable to maintain credibility with consumers. Type III eco-labels, on the other hand, are environmental statements

Table 8.1 Type-I eco-label examples

Eco-labelling system	Country	Awarding criteria
Blue Angel	Germany	It is awarded to the products meeting the criteria for reducing environmental impact in; manufacturing, recycling, packaging, use, delivery, collection and disposal stages. It also includes safety, quality, and energy consumption criteria.
Nordic Swan Eco Label/ Nordic Eco Label	Nordic countries	It is awarded to more than 60 groups of products and services, including wood fuels and wood fuel burning appliances. It provides a guarantee that the product is among the least harmful (within its product group) to the environment and health.
Austrian Eco Label	Austria	It is awarded to products, services, tourism companies, restaurants, schools and other educational institutions. It provides information about the environmental impact of consumer goods through their manufacture, use, disposal, etc.

Fig. 8.2 Nordic Swan Ecolabel [11]

that give detailed quantitative and information based on indicators. They provide standardized information based on the life cycle analysis and environmental impact of a product or service, however the actual assessment of the product is left to the consumer. Although these eco-labels are not necessarily certified, they are verified by an independent third party.

8.1.6 Rebate

Rebate is a one-time direct payment from the government to cover a specified amount of the investment cost of a renewable energy (RE) system. They are paid to the project owner upon project completion. Rebates are good at building market demand and awareness about renewable energy technologies. According to [12], a successful RE rebate program may lead to reduction in technology costs and demonstrate technological feasibility. It may also reveal market barriers, and increase the penetration of renewable energy technologies to the market. Rebates can be in the form of subsidies, low interest loans or special tax advantages. The type and the amount of incentive changes from one country or region to another. For example, in South Africa, solar water heater purchasers receive a direct rebate that depends on the type of system installed (from approximately ZAR 1900 to ZAR 4900), while in Italy, the investments in residential clean energy installations are eligible for a 36% tax rebate [5].

Lantz and Doris [12] claim that the rebates function well when they are matched with clear set of objectives that apply their strengths (market initiating nature), but they may function poorly if context-specific market factors are not considered. It is also reported that if the rebate amount is not arranged according to the existing market and policy conditions, it may not provide desired outcomes. Rebates are mostly funded by government budget, which is formed by taxes taken from the citizens. Hence, offering too high rebate levels, or providing rebates for longer than the optimal duration can cause a heavy burden on the public. This is because the amount of taxes that the citizens pay will need to increase, which will cause economic problems for the citizens. There can be several solutions to this problem. Rebates can be capped based on a maximum payment amount per project or maximum project size. For example, in Malta, for period of 2009–2010 the rebate was increased to be 66% of eligible costs up to a maximum of €460. Similarly, under the Next Generation Energy Storage program in Australia, the rebate is $825/kW up to a maximum of 30 kW [5]. CESA [13] argues that the funding should be provided in a way that it ensures the long-term continuity of the program (5–10 years). This duration allows the local markets to develop and stabilize. It is also advised to have a gradually declining level of rebates, zeroing out after 10 years.

8.1.7 Research and Development

Although primary focus is on policies that promote market transformation and deployment strategies for RE technologies, many countries sponsor a number of research and development (R&D) programs across various sectors. For example, according to [5], the United Kingdom (UK)'s leading businesses and research institutions will join forces for developing technologies to improve the economic growth and create skilled local jobs. This partnership is to be funded by a GBP 75 million joint investment from the academia, business and government. These type of partnerships or projects can boost the technological development in a country. Another good example is Belgium. Belgium's National Energy and Climate Plan (Belgium's commitment to achieve the 2030 European Union energy goals) includes several measures. One of these measures is very ambitious, as it states 3% of the Gross Domestic Product (GDP) for research and development. However, supports for R&D do not always have to be limited with one country. There are also some projects and strategic plans that cover many countries. For example, in European Union (EU), the Horizon Europe strategic plan focuses on the main orientations for the research and innovation investments over the period 2021–2024 for the EU. Horizon Europe program defines strategic areas that the EU will support research and innovation in order to help in restoring industrial leadership and opening strategic autonomy. IEA [5] reports that the financial amount for the program implementation, for the period from 1 January 2021 to 31 December 2027, is set at €86 billion.

8.1.8 Tax Credit

Tax credit scheme allows renewable energy investments to be fully or partially deducted from tax obligations [1]. It provides the investor with an annual income tax credit, based on the amount of money invested in that facility (Investment Tax Credit (ITC) or the amount of annual energy that it generates (Production Tax Credit (PTC). PTC reduces the income taxes of tax-paying owners of renewable energy projects based on the electrical output of the related renewable energy facilities. Each kilowatt-hour generated by the facility and supplied to the grid reduces the amount of income tax owed. This provides an incentive to produce more energy from renewable resources. For example, According to [14], in the United States, the PTC provides a corporate tax credit of up to 1.5 cents/kWh for electricity generated from landfill gas, open-loop biomass, and municipal solid waste resources, or up to 2.75 cents/kWh for electricity generated from geothermal, closed-loop biomass and wind resources. Although PTCs promote higher efficiency of production, they require constant monitoring of production and thus may result in high administrative costs. Additionally, PTC fails to create an incentive for the entities that do not pay taxes, such as publicly owned electric utilities and government bodies.

ITC, on the other hand, reduces income taxes for tax-paying owners based on the capital investment in renewable energy projects. Investment tax credits are earned when the capital equipment is placed into service. The federal Business Energy Investment Tax Credit (ITC) in the United States can be considered as a successful example of ITC [5]. Differing amount of ITCs are available for different technologies, ranging from 10 to 30% and decreasing/expiring over time. ITC provides an incentive to develop more capital-intensive renewable energy technologies, but do not incentivize efficient electricity production. Moreover, public utilities do not qualify for ITC, as the commercial credit cannot be claimed on public utility property.

8.1.9 Tax Reduction/Tax Exemption

Renewable energy production or investment can also be supported by tax reductions and/or exemptions. This can be done in mainly three ways:

1. The taxes for some of the RE production related expenses are exempted
2. The taxes for some of the RE production related expenses are reduced
3. Tax refund scheme is applied for the RE production related expenses

Examples to the policies mentioned in (1) can be seen all over the world. For example, based on the data in [5]; the Portuguese government provides an income tax exemption for the decentralized production of RE. In accordance with a new measure launched by the Ministry of Energy in Luxembourg, the income raised from the sale of the energy produced by small photovoltaic installations (under 10 kWp) are tax-free. Similarly, in Madagascar, with the Tax Code of 2015, the equipment for the production of RE (such as solar panels, wind or hydropower generators) is exempted from Value Added Tax (VAT).

The policies mentioned in (2) are also very popular. However, the applications may significantly vary from one country to another, in terms of the percentage deducted and the total amount available. For example, according to [5]; in Russian Federation the income tax reduction is SEK 0.60 per kWh (capped at SEK 18,000 per year) fed into the grid with a fuse size of up to 100 A. In 2004, Sweden introduced income tax reduction for house owners installing a biofuel-fired heating system in new houses. The tax reduction provided for boilers is up to 30% of the cost exceeding SEK 10,000 (the maximum relief grant: SEK 15,000 per house). Portugal, on the other hand, allows buyers of RE equipment (such as solar panels for residential applications) to benefit from a reduced VAT of 5%.

Although tax refund schemes (mentioned in (3)) are not as common as their other counterparts, there are some countries applying tax refunds for renewable energy. For example, according to the data in [5]; Ireland provides the refund of VAT paid by farmers on qualifying equipment, for the purposes of micro generation of RE electricity (solar and wind) to be used in a farm business. No matter which one of the following methods ((1), (2) or (3)) is applied, it provides direct benefits for the producer and the investor. However, the design of such tax incentives should be done very carefully, as if less taxes are taken from the citizens, the government budget decreases. If the number of citizens with tax reductions increase, the government budget decreases dramatically causing problem both for the government and for the public.

8.2 Public Finance

8.2.1 Guarantee

Guarantee (also called loan guarantee) is aimed at improving domestic lending from commercial banks for renewable energy companies and projects that have high perceived repayment risk. A loan guarantee is a kind of insurance that a debt will be repaid; a third party, such as a development bank or public lending institution takes on the risk that the borrower will not repay. A guarantee usually covers a portion of the outstanding loan principal with 50–80% being common [1]. Loan guarantees can be beneficial in many aspects [15]:

- Commercializing new technologies that may increase the performance and reduce the cost of clean energy generation
- Positioning the manufacturers to supply product for the growing global market for clean energy technologies and systems
- Near- and long-term job creation
- Contributing toward reducing emissions of various pollutants

Some examples of guarantee include [5]: the US Loan Programs Office (LPO) which provides access for up to USD 3 billion in funding through the Innovative Energy Loan Guarantee Program, and France's FOGIME, a loan guarantee fund for small and medium-sized businesses' energy sustainability investments.

8.2.2 Investment

Investment refers to the financing provided in return for an equity ownership interest in a renewable energy company or project. Typically, it is delivered as a government-managed fund that directly invests equity in projects and companies. For example, the Southern Cross Renewable Energy Fund in Australia makes equity investments in early-stage Australian RE companies to help them develop technologies, increase skills and establish international connections. Similarly, the focus of the Energy Development Corporation (EDC) in South Africa is to invest in renewable energy and alternate energy fields. EDC focuses on the following areas; hydro energy, solar energy, wind energy, bioenergy and low-smoke fuels [5].

8.2.3 Loan

Loan is financing provided to a company for a renewable energy project based on a debt (with repayment condition). Loans provided for renewable energy can be provided by the government, bank or investment authority, with lower interest rates. Loans have been used in many countries. For example, the Hungarian export bank launched a loan program through which subsidized loans having fixed term interest rates will be provided for medium and large companies that invest in RE production. In Belgium, the loans are available either through Crédal or through a Housing Fund. The loans can cover a purchase of PV panels, heat pumps and solar water heaters. Depending on the type of the fund (Crédal or Housing fund), the loan conditions change. For further details, the interested reader is referred to [5].

8.2.4 Public Procurement

The governments are major actors in economic lives of the countries. How governments spend their money is important for the emergence and diffusion of new technologies and services. Public procurement refers to the purchase by governments and state-owned enterprises of related goods and services. When applied in RE sector, the main goal of public procurement is to find and buy RE products and services with good value for money. Baron [16] lists the general stages of public procurement as follows:

- Assessment of a need for goods, services or works
- Formulation of a tender and its issuance
- Awarding and management of a contract
- Good, service or work delivery
- Monitoring

Public procurement also sets an example for the citizens of the country, as they see the installed RE systems in the public buildings or areas. The application of public procurement can change from one country to another, depending on which technology/concept needs to be supported. For example, according to [5], the City of Oslo in Norway will undertake eco-efficient procurement and will set specific climate requirements for the businesses owned by the City of Oslo. Eco-efficiency refers to creating more goods and services, but using less resources and causing less waste and pollution. In Poland, on the other hand, there is a more specific goal about alternative fuels. The Act on Electromobility and Alternative Fuels in Poland uses public procurement to support alternative fuels and introduce the concept of green public procurement (achieving the widest possible coverage of environmental issues in procurement procedures).

8.3 Regulations

8.3.1 Feed in Tariff (FIT)

In this scheme, the energy produced by the RE producer is guaranteed to be bought with a certain price for a certain amount of time. The duration depends on country (typically 5–20 years from the date the production started). A feed-in tariff (FIT) can be adjusted according to the energy market, energy infrastructure and technological conditions of a country. This makes it a popular policy tool for many countries. According to [17], FIT is currently recognized to be the most effective policy to stimulate investments in renewable energy, and it has been responsible for 45% of the global wind turbine and 75% of the global PV deployment.

FIT is a safe mechanism for the RE producers, as their electricity sales are guaranteed. Also, as it is applied for a long term, the income stays constant no matter what happens to the electricity price. This reduces the risk for the RE producers and motivates them for increasing efficiency. However, FIT also has some disadvantages. Firstly, because the electricity export price is kept constant over time, the RE producers cannot receive extra income when the electricity price increases. Also, if the FIT rate is kept constant for a long time and not reduced, the government and consumers can have economic problems due to FIT payments. Because the financing for the FIT either comes from the tax revenues or is added to the bills of the consumers. In order to eliminate such possible disadvantages, the FITs should be designed based on some factors. Some of the most commonly considered configuration factors in the design of a FIT payment structure, can be seen in Fig. 8.3. FIT level and adjustment factors, on the other hand, are illustrated in Fig. 8.4. It should be noted that Figs. 8.3 and 8.4 have been constructed by using the data provided in [17].

In designing a successful FIT mechanism, FIT configuration factors should be thoroughly considered. A careful examination would show that one policy cannot be suitable for all countries. Each country should determine the optimal setting considering their priorities, and economic and technical feasibility. It should also be kept in mind that a longer duration for FIT offers greater income to investors but comes at greater cost to consumers.

As seen from Fig. 8.4, there are also various economic factors (such as FIT level and adjustments) that should be considered in designing a FIT. The success of the program depends on the answers given to the questions asked in Fig. 8.4. Depending on which RE is needed to be promoted in a country, the policies may support some technologies or RE sources more than the other ones. Hence, there are two types of configurations:

- Flat tariff: The same unit payment is made to all RE generators, regardless of the type of installation
- Stepped tariff: Payments are differentiated by criteria such as technology, size of the installation, etc.

Most renewable energy technologies are expected to get cheaper with time, due to technological developments and increased amount of installations. In order to compensate for this, and to ensure that the support costs decrease with time, the tariffs for new applicants should reduce progressively. This is called an adjustment mechanism or degression. As seen from Fig. 8.4, there are two types of degression: Automatic and flexible. In automatic degression, scheduled tariff reductions are applied. In flexible degression however, the tariff reductions are linked to the market growth of a particular technology.

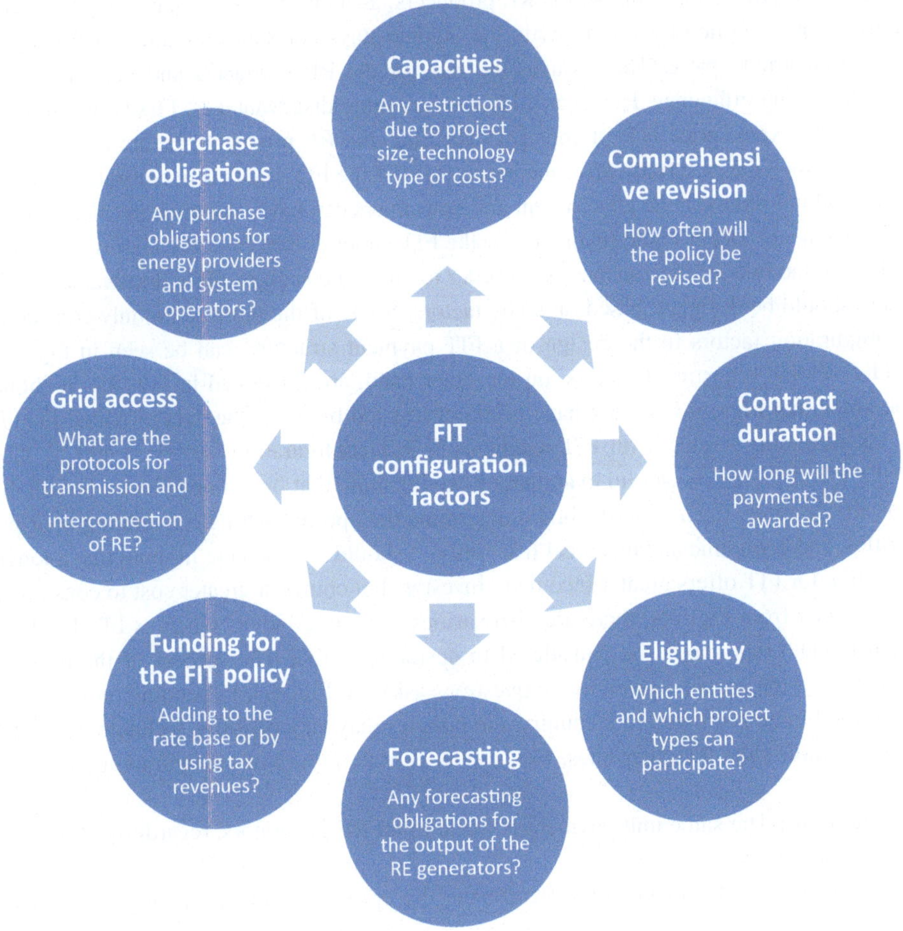

Fig. 8.3 FIT configuration factors that should be considered in FIT

8.3.2 Green Energy Purchasing

Green energy purchasing regulates the supply of voluntary renewable energy purchases by the consumers. This scheme allows residential, commercial, or industrial consumers to purchase renewable energy either directly from a utility company, from a third-party renewable energy generator or indirectly via trading of renewable energy certificates. The voluntary green energy programs usually accelerate the RE transition. In voluntary green power markets, the consumers find suppliers of RE to satisfy their commitment to RE, and the RE generators benefit from a revenue stream. Through the voluntary markets, the

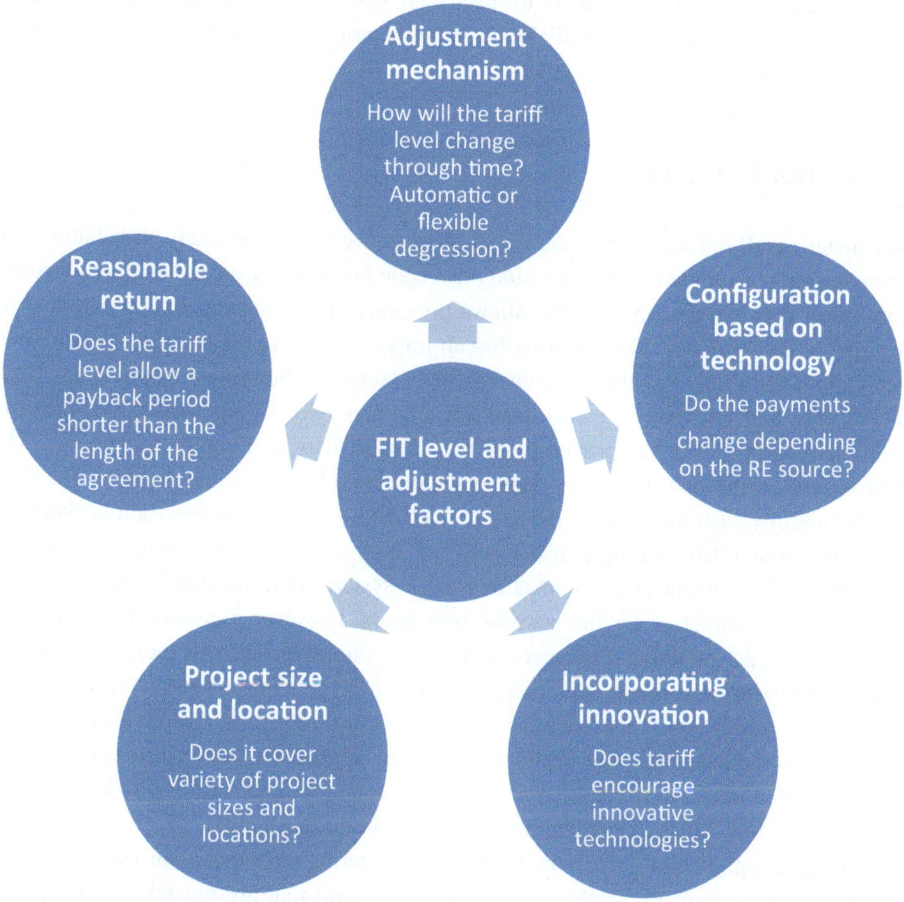

Fig. 8.4 FIT level and adjustment factors that should be considered in FIT

RE generators who cannot access the compliance markets may be able to connect directly with the buyers who desire no emissions.

In some countries or regions, green energy purchasing is applied as a green tariff. Green tariffs are voluntary utility programs which allow customers to buy energy and Renewable Energy Certificates (RECs) from a large-scale RE project, through an independent tariff. It can also be applied as a rider on the customer's electricity bill. With green tariffs, the utility companies supply consumer with up to 100% renewable power from either projects that they own or independent power producers available in the region. Green tariffs have several advantages [18]: They enable organizations to maintain existing relations with the utility, to procure RE from a project located in the same grid region as their operations, and offer price predictability and cost savings. Thus, it provides a new way for large electricity consumers to meet their sustainability and RE goals. However, usually, green tariffs are only available to large electricity customers, and may require a long term commitment.

8.3.3 Net Metering

Net metering allows a two-way electricity flow between the electricity distribution grid and consumers with their own generation (also called prosumers, as they both produce and consume). It is mainly a policy that allows prosumers to be financially compensated for the energy they produce. Net metering has an important role in deployment of distributed generation, especially via the solar energy installations in buildings. Applications differ in the way that net metering customers are compensated. Compensating net metering customers at the utility's approved retail rate of electricity is a common approach. This may be due to the fact that retail rates provide relatively high compensation. The retail rate for a utility customer reflects the total costs the utility incurs for electricity generation, transmission and distribution. If retail rates are applied for compensation, the customers can recover the upfront costs of installing a RE system more quickly, which makes net metering more attractive for the possible prosumers. However, Lawson [19] argues that compensating net metering customers at the retail rate may result in increased costs for non-net metering customers. Net metering customers generate electricity primarily for their own consumption. This reduces the amount of utility-provided electricity they need which is also beneficial for the utility as they need to produce less electricity). However, self-generation does not necessarily reduce the amount of other utility-provided services for that customer, such as maintaining the grid. When the number of prosumers in the grid increase significantly, the rates for non-net metering customers can increase so that the utility can recover the costs of maintaining the grid that are not taken from the net metering customers. This is known as a cost shift or cross-subsidy. Cross-subsidies may be affected by local factors, such as distributed generation penetration and electricity demand growth, which changes over time.

Due to the concerns about cross-subsidies, some countries/regions are exploring or adopting other options, such as adding fixed charges or minimum service fees, independent consumption and production fees, variable pricing and peer-to peer (P2P) [20]. Adding fixed charges to net metering customer's bills allow the utilities to recover the grid operation and maintenance costs. It is claimed that this increases the fairness among the consumers. However, determining a value for fixed charges that accurately reflects net metering customers' use of the grid is complex. Another application of net metering considers independent consumption and production fees. In such a case, the prosumers pay different rates for the energy consumed and produced, and, the compensation for energy production is usually lower than the cost of energy consumption. In case of applying variable pricing, the rates can change throughout the day based on the demand. Different from the others, P2P model allows non-net-metering customers and net metering customers to directly trade electricity with each other. In this model, the utility only acts as an intermediary by charging a fee for maintaining the lines connecting customers, and providing additional energy if needed. P2P is usually coupled with variable pricing of electricity.

In some countries, net metering can be applied for specific RE technologies or specific power producers. For example, with Law n°58-15, Morocco introduced a net-metering scheme for solar PV and onshore wind plants. However, according to [5], private generators may sell to the grid only up to 20% of their production. In Mauritius, the eligibility depends on the type of power producer. The main goal of their scheme is to offer the opportunity to medium-size power producers to benefit from the grid interconnection facilities by interconnecting their renewable energy installations into the grid. Similarly, the net-metering scheme in Slovenia is available for households and small businesses. The main objective of the policy however, is not to allow the export of electricity, but to support self-consumption. Hence, according to [5], if at the end of the calendar year, it is determined that more electricity is sent to the grid than acquired, the surplus will not be remunerated.

8.3.4 Premium Payment Feed in Tariff (Feed in Premium)

Premium payment feed in tariff, also called Feed in Premium (FIP), is similar to Feed in Tariff (FIT) in the sense that the energy produced by the RE producer (connected to the grid) is guaranteed to be bought for a certain amount of time. The difference is basically the type of the payment. In FIT, the total per-kWh payment is independent of the market price (constant) over a fixed period of time. In premium payment FIT, however, total payment is determined by adding a premium tariff to the spot market electricity price [1]. This premium can either be constant or varying based on a sliding scale. The differences, advantages and disadvantages of these two options are provided in Table 8.2. In the table, the eligible prosumers refer to those RE producing customers that have a premium payment feed in tariff agreement with the government or the utility company.

Sometimes, the feed in premiums can be offered as an award of the tenders or auctions. For example, in Greece, a feed-in premium is awarded to renewable and CHP (combined heat and power) plants through technology-specific tenders. In their scheme, the Ministry of Environment and Energy issues decision specifying capacities to be auctioned for each eligible technology. Similarly, in Croatia, in order to qualify for a premium tariff (offered for installations smaller than 30 kW), all renewable energy projects must win a bid in a public tender.

Table 8.2 Comparison of constant feed in premium with sliding feed in premium

	Constant feed in premium	Sliding feed in premium
Characteristics	• Total payment is equal to the market price + fixed premium • FIT premium payment level is constant over time	• When market price is higher, the premium is lower • FIT Premium payment level changes based on predetermined rules • Caps and floors (upper limits and lower limits) can be introduced
Advantages	• When the market prices are high, the eligible prosumers enjoy even higher payments • As the eligible prosumers receive high payments when the market prices are high, this can be seen as a compensation for the times that the market prices are low	• When the market prices are higher, the premium is less, and this protects the public budget and other consumers' budget • The introduction of caps and floors minimize the risks (both for the utility and for the eligible prosumers) by avoiding a large divergence between profits and losses
Disadvantages	• Paying constant premium even when the market prices are high results in higher societal costs, as the money needed for the premium may be taken from the consumer or public budget • Whenever the market prices are low, the eligible prosumers receive low payments and this may risk the profitability of the project	• Sliding premium requires a complex design in order to make sure that the market price volatility does not cause large fluctuations in RE investments

8.3.5 Priority Dispatch

Priority dispatch is a policy that mandates the transmission system operators integrating renewable energy supplies into the grid before supplies from other sources, as far as secure operation of the national electricity system permits.

WindEurope [21] claims that priority dispatch should be set according to market maturity and liberalization levels in the concerned country, but it should also take the progress in grid developments into account. When applied accordingly, priority dispatch offers many benefits. It incentivizes system operators to find solutions to minimize the amount of curtailed renewable electricity. RE curtailment refers to the reduction of power production when there is too much electricity in the grid. In order to solve this problem, the system operators may invest in system monitoring and forecasting tools. Priority dispatch also enforces the system operators and regulators to provide transparent rules on how curtailment is treated among different technologies. But, priority dispatch also has some drawbacks. EFET [22] argues that priority dispatch motivates RE producers to always run their power plant, without considering the market conditions. When RE penetration in the grid is small, this may not be a major issue. However, as the RE share in electricity mix increases, priority dispatch implies that conventional generators have to reduce their generation levels in cases of transmission congestion. Hence, there may be frequent instances in which the conventional generator has to perform a stop-start operation, and such operations are costly. This may be the reason of priority dispatch to be not-so-popular. When IEA energy policies database [5] is used to determine the number of countries that are still applying priority dispatch as a policy, it is observed that the number is comparably small. Some of the countries applying priority dispatch are as follows: Uzbekistan, El Salvador, Philippines, People's Republic of China, Kyrgyzstan, Ireland and Greece.

8.3.6 Priority or Guaranteed Access to the Network

Priority access means the first right to feed in renewable electricity in case of local grid congestion. Grid congestion happens when the electricity demand exceeds the capacity of the power grid, creating a supply demand imbalance. This leads to inefficiencies, power outages, and increased stress on the grid. It should be noted that renewable energy can also cause grid congestion. Due to their intermittent nature, wind and solar energy production can be unpredictable. If the weather conditions are very good, for example, too much sunshine or very high wind speeds, these systems may generate excess power, and the grid may struggle to accommodate it. However, when the amount of renewable energy installations increase significantly, renewable energy starts helping to relieve the stress on the grid. For example, visualize a country with several cities. All these cities would not have the same weather conditions. If the wind speeds are low in one region, they may be high in other regions of the country. Similarly, if there is high amount of cloud coverage blocking the sunlight and reducing productivity of the solar panels in one region, another region may have clear skies, causing high amount of production. This way, increasing the amount of RE installations and creating a diversity in terms of installation locations, help to relieve the congestion in the grid. Grid congestion can also be relieved by increasing investments in grid expansion and modernization, adoption of smart grid technologies, and the implementation of energy storage systems, etc. However, considering the urgency of global warming problem, the need for environmentally friendly technologies keeps increasing even more every passing day. Keeping this in mind, priority access scheme provides renewable energy supplies with unhindered access to the power grid. Some priority access programs allow renewable energy generators to able to sell and transmit their electricity at all times, whenever the source becomes available.

Although many studies refer to priority access and guaranteed access as the same thing, Alhouti [23] claims that there is an essential difference between priority access and a guaranteed access. According to [23], guaranteed access aims at ensuring RE developers with a connection to the grid, while priority access aims at providing RE developers with the priority to connect to the grid over fossil fuel-fired generators when the grid cannot accommodate all the incoming electricity. Actually, CEER [24] defines guaranteed access as follows: When the electricity from RE sources integrates into the spot market, the guarantee that all electricity sold and supported obtains access to the grid. Guaranteed access allows the use of a maximum amount of electricity from RE installations connected to the grid.

8.3.7 Renewable Portfolio Standard/Quota Obligation or Mandate

A renewable portfolio standard (RPS), also known as quota obligations, is a mandate based policy defining the minimum shares of RE sources in the energy mix of power utilities, electricity suppliers or sometimes large electricity consumers [25]. In this method, the producer, the consumer or the provider is obliged to reach a quota (which is defined by the government and increased over time). Quota is for the RE related portion of the total amount of the production, consumption, or the supply. The quota can be expressed in various ways, such as the megawatts of installed capacity, the percentage of installed capacity, the percentage of electricity produced, or the percentage of electricity sold, etc.

There is also the possibility of defining different quotas for different RE sources, with the main goal of stimulating technology diversification. Producers, consumers or providers receive a Renewable Energy Certificate (REC), which is also called Green Energy Certificate, for a fixed amount of production, consumption, or the supply (most commonly for 1 MWh). Each REC is assigned a unique serial number consisting of the tracking system's ID, date and location of generation, type of RE source, quantity, and unique identifier for tracking [25]. A company that produces/consumes/supplies more than its quota can receive more than one certificate. Then, it can sell the extra certificates (on dedicated certificate markets) to the ones that could not reach their quota. This way the ones who buy those certificates do not get penalties. The utilities are no exception to this. As stated in [26], the utilities that do not generate a high enough percentage, must purchase renewable energy certificates to make up the difference. In case of non-fulfilment, they are penalized. For example, according to [5], in Belgium, the suppliers have to give their certificates yearly to the Flemish regulator (the VREG), and if they do not comply with this obligation, they will have to pay a fine of EUR 100 for each missing certificate. The penalty rate for each missing certificate usually determines the upper level for the price of certificates. RECs only have one owner at a time. By purchasing a REC, the owner can claim the sole use of that green power. Moreover, the sale of the certificates creates a revenue stream for the operators. The revenue levels depend on the fluctuating price levels of the RECs. RECs can be traded separately from the underlying electricity generation. Hence, REC transactions create an extra source of revenue for RE generators (just like a subsidy). RECs also provide organizations with the flexibility to invest in RE even if they cannot generate it themselves, or if their utility company does not offer RE options.

Certificate allocation can have different forms, based on the desired result; uniform allocation and banded allocation [25]. Uniform allocation adopts no differentiation between different RE sources, favoring the deployment of the least-cost RE technologies. In banded allocation however, RE technologies with higher generation costs receive more than one certificate per unit of energy produced. This option has the benefit of directing the customers towards the desired or required RE sources and thus, create diversity. The number of renewable energy certificates (RECs) or green certificates received is usually established by the law. For example, in Ireland, fuel suppliers receive one certificate for each liter of biofuel placed on the market. If the biofuel is produced from materials such as biodegradable waste, residue, algae, etc., two certificates are issued. Similarly, in Romania, solar energy receives six certificates per MWh, while geothermal receives two per MWh [5].

Although RPS helps in achieving the RE policy targets in a very cost-efficient way, it also has some disadvantages. Determination of the quotas and number of certificates to be given for each technology is complex. Also, considering that the customers will try to reach their quotas by minimizing the costs, RPS may promote least-cost RE development, rather than the best. According to [27], RE producers are also exposed to the variability of the wholesale market prices, causing some degree of risk for the participants in the green energy certificate market. The risk arises due the fact that investing more than the RE required, may cause the certificate prices to remain low. This risk can be mitigated if participants are not just subject to a quota requirement, but also establish long-term contracts.

8.3.8 Tendering and Auctions

According to [28], in auctions, the price is the only criterion to be considered, while tenders may include additional criteria. Tenders are applied for an area in which the pre-feasibility study is available. Public authorities organize tenders for a determined quota of renewable energy supply capacities. Usually, a FIT is awarded to the winning bids (usually for 10–20 years). However, this support can also be in the form of feed in premiums, capacity payments, investment grants, etc. The tenders can be technology neutral or focusing on a specific RE or environmentally friendly technology. For example, Hungary had a tender to support purchase of electric cars and mopeds. In Iraq, the Ministry of Electricity of the Republic of Iraq (MOE) opened its first RE tender in 2016, for the procurement of 50 MW solar PV project. Similarly, in Turkiye, the first tender was awarded in 2017 for the construction of a 1 GW solar power plant. The solar power plant is expected to be operational for 30 years to meet the electricity demand of over 600,000 households [4].

Usually, the bidders in RE tenders have to fulfil pre-determined criteria in order to qualify for participation in the tender. Also, in many cases, bid bond guarantees are required from the bidding participants. If there is a significant delay of a project that has been selected for winning the tender, the bid bond can be retained and also penalties can be applied. These penalties can be as low as decreasing support levels or shortening the duration of support, or as high as contract termination [28]. The other criteria that may be required from the bidders or the project can be listed as follows; the technical, financial, environmental, etc. requirements related to the bidder or the RE project, and the availability of required licenses and permits for the project. For example, in Spain, the Institute for Just Transition launched a tender for 1.2 GW for RE plants. This RE project, which is expected to serve 34 municipalities affected by the closure of the coal-fired Teruel Thermal Power Plant, is to be chosen based on the following criteria; low environmental impact and positive socioeconomic impact. Similarly, the tender that was held in Turkiye for the 1GW solar plant (mentioned above) had several criteria. The tender included 65% local content, construction of a cell-manufacturing factory, employment of at least 80% local staff and a commitment to conduct research and development activities in Turkiye for a minimum of 10 years [5]. Such criteria help in developing the solar energy market, creating new job opportunities and reducing the import dependency.

Tenders have many advantages; such as increasing competition, and thus allowing high quality and low cost projects. Tenders also attract the attention of the investors like an advertisement. With tenders many people become aware of the project, the related technology and its benefits. Additionally, tenders help to reveal the true costs of RE technologies and thus prevent overcompensation. Unfortunately, the success of the tendering mechanism is highly dependent on the design of the tender and the number of participants. If there is a lack of participation to the tender, this can lead to more expensive offers, or even no suitable offer. Also, due to the high level of competition in tenders, the resulting offers can be not cost-recovering. In that case, this may negatively affect the project's future. It may even cause not being implemented. Moreover, if the tenders are not organized regularly, this would not support continuous RE market development. The interested reader is referred to [29, 30] for more detailed information about tendering. Renewable Energy Sources Work Stream of Electricity Working Group [29] describes the tendering procedures for RE in Europe, while [30] provides a comprehensive guide for designing RE auctions.

8.4 Choosing the Most Suitable Renewable Energy Policy

As seen from the previous subsections, there are many different types of RE policies. Not all the policies are appropriate for all the countries. In that case, which policies or support mechanisms are the best for a specific country? How can this be determined? This section

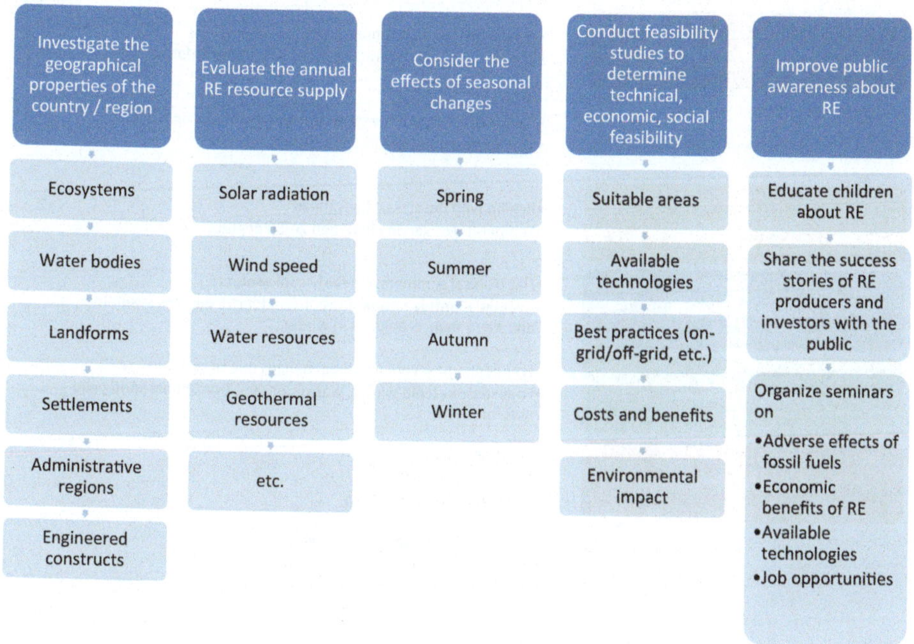

Fig. 8.5 Steps for determining the best RE policies to be applied

will provide answers to these questions. In order to determine the best RE policies to be applied, one should follow the steps shown in Fig. 8.5.

Geographical properties of a country are important for deciding where the systems can be installed and where they should not be installed. After this analysis, the type of renewable to be supported should be decided. This can be done by analyzing the RE potential of the country. The RE potential mainly depends on the climate zone in which the country is located. Figure 8.6, provides the characteristics of different climate zones. This figure is constructed by using the data provided in [31]. The degrees mentioned in the figure are in Celsius.

For example, wind has a direct relation with the climate zones. Winds occur on a various range of scales, from thunderstorm flows, to local breezes (generated by heating of land surfaces), and to global winds (resulting from the difference in solar energy absorption between the climate zones on Earth) [32]. While some regions receive too much rainfall, some others have long and dry seasons with high amount of sunshine. Solar PV systems require clear skies with high amount of solar radiation. However, the temperatures should not be so high so that the efficiency of the solar panels would not reduce. However, in those locations with high temperatures, solar should not simply be eliminated as it can be used for solar water heating purposes or concentrated solar power (CSP). Similarly, in regions receiving high amount of rainfall or having a considerable amount of ice or

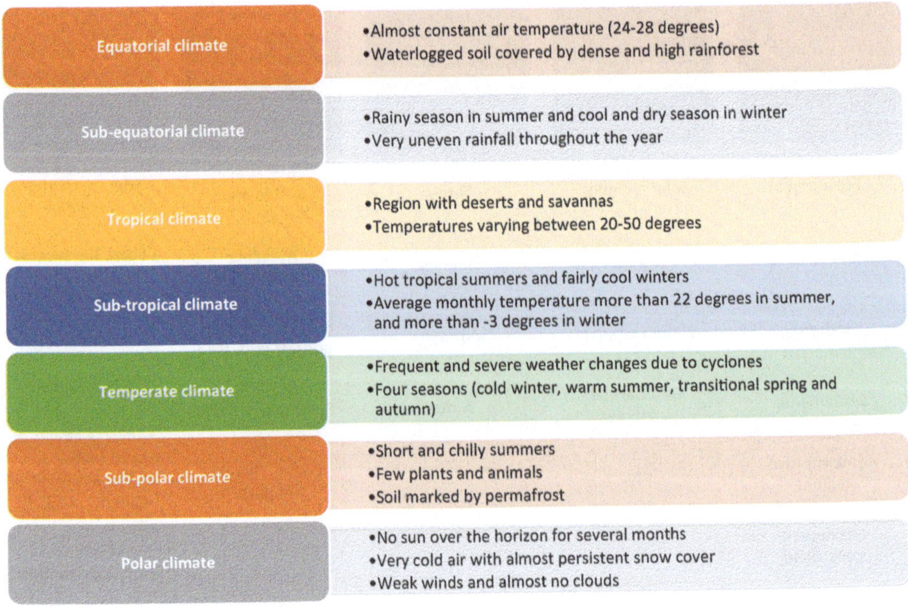

Fig. 8.6 Characteristics of different climate zones

snow cover, hydropower potential increases significantly. In relevance to that, those areas receiving high amount of rainfall usually have high amount of bioenergy potential. For example, Cyprus, an island located in Mediterranean, is characterized by mild rainy winters, occasional droughts, and long and hot summers. This makes it a favorable place for solar energy but not a favorable place for hydropower. The climate of the United States, however, is highly diverse. It ranges from tropical conditions in south Florida and Hawaii, to arctic and alpine conditions in Alaska and across the Rocky Mountains. This allows the country to benefit from many different types of renewable sources [33].

However, it is not only the climate zones that affect the RE potential of the countries, geographical properties of a country can also have an effect on its RE potential. Moreover, the seasonal changes in wind or rainfall patterns can have a significant impact on bioenergy and wind power production, and hydropower generation. Sometimes, even if there is RE potential somewhere, it may not be feasible to install a system there, due to; political problems, economic issues, lack of technology or NIMBYism. NIMBYism, in which NIMBY means Not in My Backyard, is a term used for local opposition to projects designed for public interest and accused to have a negative impact on its surroundings. Some of the NIMBY themes regarding solar and wind energy are; noise, land-use implications, wildlife concerns, and aesthetic concerns. Hence, deciding the correct RE policies is a complex task. Even after the RE policy is selected and adopted, its success depends on the attention and compliance of the possible RE producers and investors to those policies. In order to increase the amount of investors, people should be trained about RE, so that they start benefiting from the policies, and this in turn increases the share of RE in the country.

8.5 Exercises

1. What are the possible effects of capacity building on RE development?
2. What is the major difference between carbon tax and energy tax?
3. What is the major difference between grant and rebate?
4. What is green labelling? Why is it used for?
5. What are the required design/implementation principles for effective eco-labelling?
6. What is the major difference between investment tax credit and production tax credit?
7. What are the possible drawbacks of using tax reduction or tax exemption for supporting RE?
8. What are the general stages of public procurement?
9. What is the major difference between feed in tariff and feed in premium?
10. What are the configuration factors of feed in tariff?
11. What are the factors affecting FIT level adjustments?
12. What is nct metering?
13. What are the possible advantages of sliding feed-in-premium over constant feed in premium?
14. What are renewable energy certificates? For which purpose are they used for?
15. How can one choose the most suitable renewable energy policy for a country or a region?

References

1. IRENA. (2012). *Evaluatıng policies in support of the deployment of renewable power.* IRENA Polıcy Brıef. International Renewable Energy Agency (IRENA). Retrieved April 2024, from https://www.irena.org/-/media/Files/IRENA/Agency/Publication/2012/Evaluating_policies_in_support_of_the_deployment_of_renewable_power.pdf?rev=524a56e7b57145e5b988d2276844b631
2. IRENA Secretariat. (2012). *Approach paper for the IRENA capacity building strategy.* IRENA. Retrieved April 2024, from https://www.irena.org/-/media/Files/IRENA/Agency/Events/2012/Jun/6/IRENA_CB_Strateg_Approach_Paper.pdf?la=en&hash=0DF2A1DBC1A6C95AAB6244FFD7A751AC62DEC62E
3. Mungai, E. (2023). *The need for capacity building for organizations when it comes to sustainability matters.* Africa Sustainability Matters (ASM). Retrieved April 2024, from https://africasustainabilitymatters.com/the-need-for-capacity-building-for-organizations-when-it-comes-to-sustainability-matters/

4. OECD. (2021). *Taxing energy use for sustainable development.* OECD. Retrieved April 2024, from https://www.oecd.org/tax/tax-policy/taxing-energy-use-for-sustainable-development.pdf
5. IEA. (2024). *Policies database.* IEA. Retrieved April 2024, from https://www.iea.org/policies
6. Boucher, K., Guerra, J., & Watkins, B. (2010). *Auburn, Massachusetts wind feasibility study.* Project report. Worcester Polytechnic Institute.
7. Boström, M., & Klintman, M. (2008). *Eco-standards, product labelling and green.* Palgrave Macmillan.
8. UNESCAP. (2024). *Eco-labelling. Low carbon green growth roadmap for Asia and the Pacific: Fact sheet.* Retrieved April 2024, from https://www.unescap.org/sites/default/d8files/2021-11/18.%20FS-Eco-labelling.pdf
9. Confederación Nacional De La Construcción Associazione Nazionale De Costruttori Edili. (2018). *Module 2: Certification and labelling.* Erasmus+. Retrieved April 2024, from https://ec.europa.eu/programmes/erasmus-plus/project-result-content/2232f669-b515-4328-ab19-f764af1e819f/R7_Learning_Outcomes_M2__Certification_and_Labelling_EN.pdf
10. Bürger, V. (2007). *Green power labelling.* Öko-Institut. Retrieved April 2024, from https://www.oeko.de/oekodoc/1480/2007-230-en.pdf
11. Nordic Swan Ecolabel. (2024). *Regulations, guidelines and logos.* Nordic Swan Ecolabel. Retrieved April 2024, from https://www.nordic-swan-ecolabel.org/how-to-apply/regulations-guidelines-logos/
12. Lantz, E., & Doris, E. (2009). *State clean energy practices: Renewable energy rebates.* Technical report. National Renewable Energy Laboratory. Retrieved April 2024, from https://www.nrel.gov/docs/fy09osti/45039.pdf
13. CESA. (2009). *Developing an effective state clean energy program: Renewable energy incentives.* Briefing paper. Clean Energy States Alliance. Retrieved April 2024, from https://www.cesa.org/wp-content/uploads/CESA-renewable-energy-incentives-mar09.pdf
14. EPA. (2024). *Renewable electricity production tax credit information.* United States Environmental Protection Agency. Retrieved April 2024, from https://www.epa.gov/lmop/renewable-electricity-production-tax-credit-information
15. CRS. (2012). *Loan guarantees for clean energy technologies: Goals, concerns, and policy options.* Congressional Research Service. Retrieved April 2024, from https://crsreports.congress.gov/product/pdf/R/R42152
16. Baron, R. (2016). *The role of public procurement in low-carbon innovation.* OECD. Retrieved April 2024, from https://www.oecd.org/sd-roundtable/papersandpublications/The%20Role%20of%20Public%20Procurement%20in%20Low-carbon%20Innovation.pdf
17. UNESCAP. (2024). *Feed-in tariff. Low carbon green growth roadmap for Asia and the Pacific: Fact sheet.* Retrieved April 2024, from https://www.unescap.org/sites/default/files/26.%20FS-Feed-In-Tariff.pdf
18. EPA. (2024). *Utility green tariffs.* United States Environmental Protection Agency. Retrieved April 2024, from https://www.epa.gov/green-power-markets/utility-green-tariffs
19. Lawson, A. J. (2019). *Net metering: In brief.* Congressional Research Service. Retrieved April 2024, from https://crsreports.congress.gov/product/pdf/R/R46010
20. Most Policy İnitiative. (2024). *Net metering. Most policy initiative.* Retrieved April 2024, from https://mostpolicyinitiative.org/wp-content/uploads/2022/02/Net-Metering-1.pdf
21. WindEurope. (2016). *WindEurope views on curtailment of wind power and its links to priority dispatch.* WindEurope. Retrieved April 2024, from https://windeurope.org/wp-content/uploads/files/policy/position-papers/WindEurope-Priority-Dispatch-and Curtailment.pdf
22. EFET. (2012). *EFET response to CEER consultation on implications of non-harmonised renewable support schemes.* EFET (European Federation of Energy Traders). Retrieved April

2024, from https://data.efet.org/Files/Documents/Emissions%20and%20RES/RES-E/EFET_response_CEER_Consultation_06012012.pdf
23. Alhouti, A. (2024). *Deployment of solar energy in Saudi Arabia: A case study*. The George Washington University Law School. Retrieved April 2024, from https://gwujeel.files.wordpress.com/2013/09/johara-alhouti.pdf
24. CEER. (2013). *Status review of renewable and energy efficiency support schemes in Europe*. Council of European Energy Regulators ASBL. Retrieved April 2024, from https://www.ceer.eu/documents/104400/-/-/fb65b156-71a6-b717-a3ec-585b138aa3ae
25. Energypedia. (2024). *Renewable energy quota and certificate schemes*. Energypedia. Retrieved April 2024, from https://energypedia.info/wiki/Renewable_Energy_Quota_and_Certificate_Schemes
26. Abbas, A., Price, C., & Wenning T. (2022). *Renewable energy certificates*. ORNL/SPR. Retrieved April 2024, from https://energyefficiency.ornl.gov/wp-content/uploads/2022/08/Better-Plants-Renewable-Energy-Certificates-Overview_Aug-2022.pdf
27. Santana, R. B. (2017). *Impact of an optimum renewable portfolio standard in the system adequacy and its effect on the wholesale electricity market: Dominican Republic*. Retrieved April 2024, from https://repositorio.comillas.edu/jspui/bitstream/11531/24582/1/TFM000829.pdf
28. Energypedia. (2024). *Renewable energy tendering schemes*. Energypedia. Retrieved April 2024, from https://energypedia.info/wiki/Renewable_Energy_Tendering_Schemes
29. Renewable Energy Sources Work Stream of Electricity Working Group. (2023). *CEER report on tendering procedures for renewable energy sources in Europe*. Council of European Energy Regulators ASBL. Retrieved April 2024, from https://www.ceer.eu/documents/104400/-/-/de58ad59-2089-979e-12b4-22f5f250a9a6
30. IRENA and CEM. (2015). *Renewable energy auctions—A guide to design*.
31. UNDP. (2024). 1.2. *Types of climate and climate zones*. United Nations Development Programme. Retrieved April 2024, from https://climate-box.com/textbooks/the-problem-of-climate-change/2-2-effects-on-plants-and-animals/
32. Wikipedia. (2024). *Wind*. Wikipedia. Retrieved April 2024, from https://en.wikipedia.org/wiki/Wind
33. Climate Change Knowledge Portal. (2024). *Climatology. Climate change knowledge portal*. World Bank Group. Retrieved April 2024, from https://climateknowledgeportal.worldbank.org/country/united-states/climate-data-historical

9 Renewable Energy Policies in Heating and Cooling Sector

As stated in [1], modern renewable heat covers direct and indirect (for example through district heating) final consumption of; bioenergy, solar thermal energy, geothermal energy, and renewable electricity for heat (based on an estimate of the amount of electricity used for heat production). Unfortunately, heat and transport lag behind electricity in terms of RE uptake, despite accounting for 77.3% of the global final energy supply [2]. Therefore, there is a need to support RE adoption with the use of policies. The general renewable energy policies have been explained in the previous section. This chapter is dedicated to discussing the adoption of such RE policies for heating and cooling purposes. At this point, it should be noted that although there may be many different policy types used in promotion of RE for heating and cooling, the RE policy types covered in this chapter are mainly the policy instruments listed in [3] for promoting the use of renewables to produce heat. It should be noted that this work is partially based on 'Renewable Energy Policies in a Time of Transition' developed by IRENA, OECD/IEA and REN21 [3] but the resulting work has been prepared by Neyre Tekbıyık Ersoy and does not necessarily reflect the views of IRENA, OECD/IEA nor REN21. Neither IRENA, OECD/IEA nor REN21 accepts any responsibility or liability for this work. The same is valid for the next two chapters written on RE policies in transport sector and RE policies in power sector.

In this chapter, the policies are presented according to the categorization provided in this book. Figure 9.1 illustrates the related renewable energy policies according to their categories; incentives, education and research, public finance and regulations.

N. Tekbıyık Ersoy, *Energy Efficiency and Renewable Energy Policies*, Synthesis Lectures on Renewable Energy Technologies, https://doi.org/10.1007/978-3-031-64305-7_9

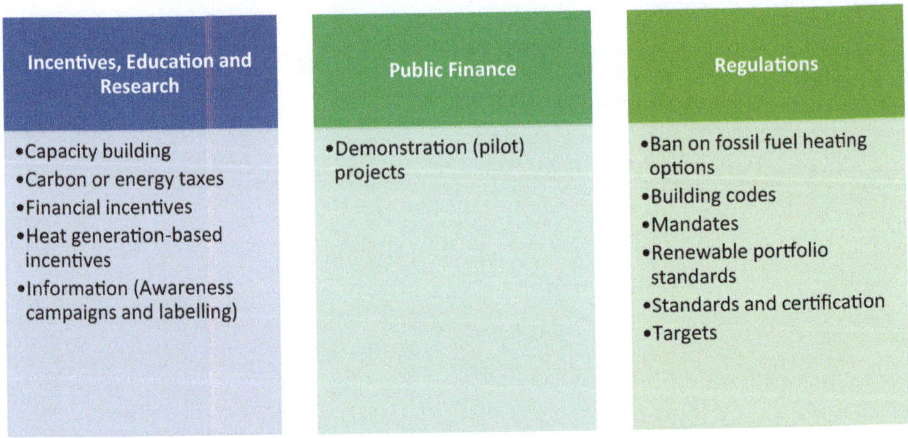

Fig. 9.1 Renewable energy policies adopted in heating and cooling sector

9.1 Incentives, Education and Research

Let's start the discussion with incentives. As seen from the figure, incentives for heating and cooling can be of various types; carbon or energy taxes, financial incentives, and heat generation based incentives. Carbon taxes and energy taxes serve as important price signals. Price signals are changes in the price of a service or a good that are intended to deliver information to consumers/producers in order to influence their behavior. Therefore, especially carbon taxes can motivate people to invest in renewable energy or prefer using renewable energy sources rather than fossil fuels, for heating and cooling purposes. Actually, according to [3], energy taxes have been an important driver of the shift to renewable fuels in Nordic countries.

Carbon taxes also deal with externalities. In direct proportion to the meaning of externalities, people using fossil fuels directly or indirectly contribute to the global warming, but they may not be the ones suffering the most from it. Carbon taxes make sure that those who cause emissions somehow pay penalties for their choices. For example, in Mexico, within the tax law of the country, the carbon tax of approximately 3 USD/ton of carbon on imported or locally produced fuels is applied. The related carbon tax is updated annually. Similarly, in Ireland, the carbon tax is updated annually, and the tax applies to the emissions from burning petroleum, kerosene, fuel oil, auto-diesel, marked gas oil, liquid petroleum gas (LPG), natural gas and solid fuels (such as peat and coal). According to [4], the tax was raised to €26 from May 2020, which shows the dedication of the country for decreasing the amount of emissions. However, it should be noted that this is not the highest carbon tax in the world. According to the same reference, Sweden has the highest CO_2 tax in the world, €114... However, this success did not happen over a day. With the gradual increase of the tax level, Sweden provided enough time for the households and businesses to adapt, and this has improved the political feasibility of the increases and it created an atmosphere of trust. Taking such successful practices as an example, any country that will introduce carbon prices, can start with a low price and increase it over time. This allows the gradual decrease of fossil fuel usage for those countries that have difficulty in making the transition. However, carbon taxes and energy taxes are politically difficult to implement. Moreover, in such developing countries that there is a need for certain industries to continue their operation for further development of the country, exemptions from taxes can be granted. This, in turn, can make the carbon taxes less effective.

Financial incentives, such as grants, tax credits and investment subsidies can also be used for promoting RE in heating and cooling. Such incentives may come in many different forms. For example, the Belgian government enacted a temporary VAT reduction in 2022 to the delivery and installation of solar water heaters and heat pumps. While some countries adopt one type of incentive, some others may adopt several in order to create diversity in support and increase the speed of adoption of RE technologies. For example, according to [4], Chile has three mechanisms to promote installing solar water heaters for residential water heating; a tax exemption for solar water heater installation in new housing, a subsidy to incorporate solar water heating into housing reconstruction programs, and the Ministry of Housing and Urban Development (MINVU)'s family heritage protection program for existing public housing. It should be noted that such incentives are not only applied for heating. For example, as elaborated in [4]; in Italy, the Conto Termico 2.0 provides financial incentives on the capital costs of the initiatives, that are paid annually for a variable period of 2–5 years, based on the type of improvement implemented, technology applied and its size. Some of the eligible technologies are; heat pumps, biomass boilers, and solar thermal systems (including those based on the solar cooling technology).

It is clear that the share of renewables in heating and cooling is much lower than the one in electricity. According to [2], renewable heat accounted for 11.5% of total heat demand in 2020 (excluding traditional biomass which accounts for 13.1%), while renewable electricity accounted for an estimated 29.9% of the total global electricity production in 2022. Hence, financial incentives can promote the use of renewable heat and can improve the competitiveness of renewable heat compared to fossil fuels. It can also be helpful in addressing the barrier of higher capital costs in RE based heating. However, support levels may change due to shifting political priorities (especially in fossil fuel-rich countries or countries having high energy dependence on fossil fuel-exporting countries), and this may affect the long term effects of the incentives.

Heat generation based incentives are similar to feed-in-tariffs. Hence, they provide support over a long time. According to [3], in the United Kingdom, there is a renewable heat incentive (RHI), in which the owner of the renewable heat equipment receives a tariff per kilowatt hour generated, with the scale depending on the technology used. The related RHI payments for non-domestic generators are awarded for period of 20 years, and the eligible technologies are [4]; aerothermal energy, hydrothermal energy, biogas, biomass, solar thermal and geothermal installations.

There are many projects and programs supported by development aids, which can be multilateral (in which all systems directly linked) or bilateral (in which the systems are directly and indirectly linked) in nature. Such aids usually support either capacity building in the supply chain or the take-up of clean cooking solutions. Especially in those regions where people lack access to clean cooking facilities, capacity building is an important scheme to ensure that local enterprises can shift to clean cooking. According to [3], the Global Alliance for Clean Cookstoves (GACC) operates a capacity building facility for helping local enterprises to bring their operations to scale, and achieve commercial viability and leverage private investment. Although capacity building is important for supply chains, unfortunately, it is unlikely to result in much deployment on its own as there may be other barriers to the development, such as economic, social, technical, etc. feasibility.

Awareness campaigns and labeling are also important for supporting RE in heating and cooling, especially in countries competing with extensive individual natural gas heating. The characteristics of such countries are [3]; comparably high demand for space heating, use of individual boilers and majority of residential and commercial buildings connecting to the natural gas grid, and relatively low gas prices. Due to the low penetration rates of renewable heat in those countries, policies that promote public awareness and confidence (such as certification) are necessary. Some countries, such as Nepal, want to increase awareness about access to reliable, clean and suitable energy in the rural areas. Among the several objectives of the Rural Energy Policy of Nepal, one stands out as it is related with heating. According to the related policy; RE public awareness/promotional programs will be put in place in order to increase the use of solar cookers and improved cook stoves. Sometimes, countries may want to increase awareness in specific technologies. For example, Canada has the Northwest Territories Solar Strategy which aims to increase education and awareness about solar energy technologies for a diverse set of people and organizations, such as; businesses, communities, government departments, and residents [4].

Although information based policies are essential for creating awareness about RE options and their costs/benefits, they may not be as effective as expected owing to acceptability problems. IRENA, OECD/IEA and REN21 [3] claims that such policies are most effective when done as a part of personalized energy advice. However, that is expensive to deliver.

9.2 Public Finance

Demonstration (pilot) projects can also be beneficial in support or RE heating as it allows testing related RE technologies for local suitability and possible outcomes. For example, Poland has a pilot program that aims to reduce the negative impact of district heating companies on the environment by supporting investment projects. In United States California, there is a solar hot water heating pilot program, which is one of the components of California Solar Initiative. This initiative aims at putting solar on a million roofs in the state of California. Similarly, Pilot Bioheat Boiler Deployment Program in Ireland aims at promoting biomass boilers for space heating by providing capital grant support for the installation of some biomass boilers, for small industrial sites and large buildings [4]. Unfortunately, usually, demonstration or pilot projects are unlikely to result in widespread deployment on their own. Hence, they should be supported with other measures suitable to the country or region.

9.3 Regulations

Many countries have applied regulatory measures for supporting renewable heating and cooling. For example, Denmark, has adopted bans on fossil fuels for space heating. According to [3], since 2013, oil or gas heating has not been allowed in new buildings in Denmark. Moreover, since 2016, new heating oil unit installations have been prohibited in existing buildings in areas supplied by district heating or natural gas. Similarly, Finland is planning to ban the coal energy use by 2029, and to have emission-free heat production by 2030 [4]. Such bans on fossil fuels provide greater certainty of success due to their mandatory nature. However, if a specific fuel type is banned, then there should be an alternative to it so that the consumers can easily access. Otherwise, such bans may not yield to successful results.

Building codes also promote using renewables for heating and cooling purposes. For example, in United Arab Emirates, the emirate of Dubai requires that new buildings meet to 75% of their water heating requirements by solar energy. If there is a swimming pool in the building, then, 50% of its water heating requirements must be met by solar energy [4]. Similarly, according to [5], in United States, new federal buildings and major renovations of the federal buildings must meet 30% of their hot water demand from solar water heaters, if cost-effective. Building codes also provide an opportunity to align energy efficiency with renewable heat requirements. Such as requiring the houses to have a certain amount of insulation, or double glazed windows, as well as having solar air/water heaters, etc.

Mandates are sometimes similar to building codes due to their enforcing nature. However, they are not restricted to the buildings only. For example, the law in Bolivarian Republic of Venezuela mandates ministries to use RE for thermal purposes. Antigua and

Barbuda mandates the installation of solar water heating systems in the tourism sector. There can also be technology specific mandates, such as the solar thermal mandate in Uruguay. According to [4], in Uruguay, all new construction and renovations of public buildings, health/sports facilities, and hotels, where hot water accounts for more than 20% of the building's energy consumption, must obtain minimum 50% of water heating energy from solar thermal energy. Although mandates and building codes provide a greater certainty for increasing deployment, they are mostly applied to the new buildings, and therefore do not cover the overall heat demand.

Renewable portfolio standards (RPSs) are similar to mandates, as mentioned earlier. This makes them provide certainty over RE deployment levels in heating and cooling. However these standards are usually much less ambitious for heating than they are for electricity, lowering down their effect on development. Standards and certifications are more common as they ensure the reliability of a technology. For example, in Australia, a standard on solar air conditioning, called AS 5389 covers a methodology for calculating the heating or cooling loads of buildings under a variety of climatic conditions [4]. Sometimes, standards can also be used as benchmarks. For example, in Ireland, under the House of Tomorrow program, developers, who designed buildings to consume 40% less energy for water and space heating than the current building regulations minimum standards, could be awarded EUR 8,000 per dwelling, with a minimum and maximum number of dwellings to be supported to be determined by the scheme.

Finally, targets can be enacted in order to improve RE adoption in heating or cooling. The targets do not even have to be about RE in order to support RE. For example, The Dutch government aims to reduce the role of gas for heating buildings as a part of its long-term targets for reducing CO_2 emissions. This is expected to indirectly support RE, as there would be a need for an alternative when gas cannot be used. Similarly, Finland announced that it will phase out all coal by 2030. Considering that coal currently accounts for 30% of heat consumption, this is an important step for the heating sector. However, some countries' targets are even more ambitious. For example, in Norway, 43% of heat consumption should be met by renewable sources. Similarly, in Denmark, the main objective of the Danish 2050 Energy Strategy is achieving 100% independence from fossil fuels in the national energy mix, by 2050.

Although targets are not usually effective on their own, they provide a clear direction of what needs to be achieved in the future and when. In order to make sure that the targets can be achieved, supportive policies are needed.

9.4 Exercises

1. What is the share of renewable heat in total heat demand? How is that share when compared to the share of RE in electricity?
2. What are the types of regulatory policies applied in heating and cooling sector?

3. How does the application of tax related incentives change from one country to another?
4. What is the effect of carbon taxes on RE development for heating purposes?
5. Which kind of building codes are applied in order to promote using renewables for heating and cooling purposes?
6. Which countries have more ambitious RE targets in heating and cooling sector?
7. Which RE policies should be adopted in those countries with low penetration rates of renewable heat?

References

1. IEA. (2020). *Renewables 2020: Analysis and forecast to 2025.* International Energy Agency. Retrieved April 2024, from https://iea.blob.core.windows.net/assets/1a24f1fe-c971-4c25-964a-57d0f31eb97b/Renewables_2020-PDF.pdf
2. REN21. (2024). *Global overview.* REN21. Retrieved April 2024, from https://www.ren21.net/gsr-2023/modules/global_overview/
3. IRENA, OECD/IEA and REN21. (2018). *Renewable energy policies in a time of transition.* IRENA, OECD/IEA and REN21. Retrieved April 2024, from https://www.irena.org/-/media/Files/IRENA/Agency/Publication/2018/Apr/IRENA_IEA_REN21_Policies_2018.pdf
4. IEA. (2024). *Policies database.* IEA. Retrieved April 2024, from https://www.iea.org/policies
5. Kaufmann, J., Hand, J., & Halverson, M. (2011). *Integrating renewable energy requirements into building energy codes.* Pacific Northwest National Laboratory. Retrieved April 2024, from https://www.pnnl.gov/main/publications/external/technical_reports/pnnl-20442.pdf

10 Renewable Energy Policies in Transportation Sector

According to [1], of the three sectors, transport has the lowest penetration of renewables. In transport demand, renewables grow from 3.4% in 2017 to just 3.8% in 2023. Considering that the primary fuels used in transportation sector cause significant amount of greenhouse gas emissions, it is important to find ways of lowering down those emissions. Electrification of transportation sector can be the key in accomplishing that task. Therefore, many RE policies focus on electric vehicles. However, the policies are not only concerned with electric vehicles. This chapter focuses on different types of RE policies adopted in transportation sector. Although there may be many different policy types used in promotion of RE in transportation sector, the RE policy types covered in this section are mostly the policy instruments listed in [2] for promoting the use of renewables in transportation. The categorization adopted in this book is also followed in this chapter. Figure 10.1 illustrates the related renewable energy policies according to their categories; incentives, education and research, public finance and regulations.

10.1 Incentives, Education and Research

Due to the high dependence of the transport sector on fossil fuels, removal of fossil fuel subsidies is a necessity to decarbonize the transport sector. Fossil fuel subsidies can be considered as the government actions that decrease the cost of exploration and production of fossil fuels, and their price to the consumers. Milicevic [3] argues that the existence of fossil fuel subsidies endangers the achievement of the following sustainable development goals (SDGs); SDG7, SDG12 and SDG13. According to [4], SDG7 focuses on ensuring access to affordable, reliable, sustainable and modern energy for all. SDG12's

N. Tekbıyık Ersoy, *Energy Efficiency and Renewable Energy Policies*, Synthesis Lectures on Renewable Energy Technologies, https://doi.org/10.1007/978-3-031-64305-7_10

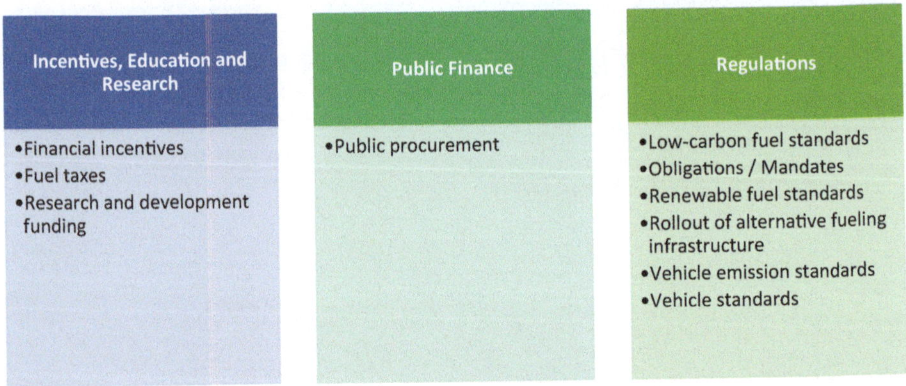

Fig. 10.1 Renewable energy policies adopted in transportation sector

main concern is to ensure sustainable consumption and production patterns. SDG13, on the other hand, is about taking urgent action to combat climate change and its impacts. Fossil fuel subsidies make the transition to zero carbon less likely, as supporting fossil fuels means increasing the use of fossil fuels, which means increasing its combustion, which in turn means more emissions. Hence, fossil fuel subsidies should be phased out and instead, the renewable energy subsidies/incentives should be applied more widely.

In transportation sector, the financial incentives are mostly used for making electric vehicles competitive with conventional vehicles. The incentives can be in the form of tax breaks, exemptions or rebates in favor of low emissions. According to [2], some countries such as China, Brazil, 20 members of the European Union (EU) and South Africa collect differentiated taxes on vehicle registration and/or circulation based on their fuel economy or emissions performance. This can also be in the form of tax exemption. For example, in China, new energy saving vehicles (such as fuel cell vehicles, battery electric vehicles, plug-in hybrid electric vehicles, etc.) are entirely exempted from vehicle and vessel tax. Switzerland also applies tax exemption with some conditions; biofuels must generate at least 40% less greenhouse gas (GHG) emissions compared to the life cycle emissions of fossil fuel, and the cultivation of raw materials must not endanger biological diversity and tropical forests [5]. In addition to the type of incentives mentioned above, fuel taxes can also be used in support of RE in transportation sector. Fuel taxes are usually based on the life-cycle of GHG emissions for fuel. Although they increase the competitiveness of renewable options with fossil fuels, some people may argue that determining life cycle emissions is complex, time consuming and therefore open to debate.

The technologies that have long-term market potential but high investment risk (due to potential project failure) need to be supported with research and development (R&D) funding. Good examples to this can be biofuels and Power-to-X, considering their capital intensive nature. Some countries like the United States and Australia support the development of advanced biofuels with grants for research and development (R&D) [2]. Some countries prefer supporting Power-to-X rather than biofuels. Power-to-X refers to conversion technologies that allow for the decoupling of power from the electricity sector for use in other sectors (such as transport or chemicals). For example, Finland's recovery plan includes the R&D sub-area. The investments are to be directed to R&D projects focusing on the priorities set by the first pillar of the plan (green transition). One of these priorities is the new uses for low-carbon hydrogen (i.e. Power-to-X) [5].

Some countries also provide financial incentives for R&D. According to [2], those countries implement financial incentives to support R&D into new technologies that integrate RE directly into vehicles, such as what has been done by the government of Uganda. The government has supported the development of a solar-powered bus that integrates photovoltaics on the roof. Sometimes, support can be not just a financial incentive but also a collaborative effort. In Netherlands for example, several companies, universities, grid operators and Dutch municipalities have joined the Living Lab Smart Charging platform supported by the national government [2]. The platform has a very ambitious goal of ensuring that solar energy and wind energy power all electric vehicles in the country. The lab is installing thousands of new smart-charging ready stations for research and testing in order to develop international standards.

10.2 Public Finance

The adoption of RE in electricity is much more than other sectors, such as heating and cooling and transportation. That's why making such sectors more electricity friendly can be the key in increasing the integration of RE in that sectors. Public procurement policies including electric vehicles can serve as a starting point for increasing RE development in transportation sector. For example, Poland aims at boosting low carbon transport via amendments to existing legal acts on electromobility in order to increase the amount of green vehicles in public procurement [5]. Sometimes, public procurement policies can be much more than just about electric vehicles. The countries may require considering environmental, social, labour, etc. laws in public procurement. Unfortunately, public procurement only accounts for a limited part of the demand and therefore, must be complemented with other measures/policies in order to broaden RE penetration.

10.3 Regulations

There are many different types of regulatory measures that can be used in support of RE in transportation sector. These could be summarized as, but are not limited to; low-carbon fuel standards, obligations or mandates, renewable fuel standards, rollout of alternative fueling infrastructure, vehicle emission standards, and vehicle standards. As it may be seen from this list, most of these policies are either standards or mandates.

Low carbon fuel standards (LCFSs) are programs that aim to reduce carbon dioxide emissions and other pollutants from transportation fuels. Policy makers are attracted by LCFSs, as such standards send a clear policy signal to investors that long-term solutions are needed for cost-competitive and lower-carbon transportation fuels. LCFSs are technology neutral and they allow a variety of different alternative fuels (gaseous fuels, biofuels, electricity, etc.) to participate as long as they have small carbon footprints. They have an immediate effect by improving the emissions intensity of fuels used today. Moreover, they encourage clean innovation and speed up the transition to cleaner fuels for future. According to [6], California's LCFS has been successful in incentivizing the creation of a large domestic market for low carbon fuels that emit 65–80% less carbon dioxide per unit of energy than petroleum fuels. Some LCFSs, like the one in Canada, allow suppliers flexibility in reaching the standard, via credit trading. In that case, the suppliers of alternative low-carbon fuels can earn credits. Any fuel supplier or importer who supplies fuel that falls below the maximum carbon intensity for that year generates credits, which can be saved for a future year or sold [7].

In contrast to an LCFS, a Renewable Fuel Standard (RFS) requires a certain volume of renewable fuel content in diesel and gasoline fuel sales, but it does not cover alternative fuel sources as an LCFS does. For example, in United States, the RFS is a federal program that requires the energy companies to blend billions of gallons of corn ethanol and other biofuels into the gasoline supply each year [8]. According to [7], although there is an incentive to reach the required standard for RFS, unlike LCFS there is no incentive for additional emission reductions. In LCFS, the lower the carbon intensity of the fuel, the greater the reward. Hence, when used in isolation, RFS may result in more costly emission reductions than an LCFS. The reader interested in getting more information about the distinctions between these two types of standards is referred to [9].

Many countries also adopt obligations/mandates in order to facilitate the adoption of RE in transportation sector. These obligations can be in the form of mandating the share of renewables in fuel or zero-emission vehicle mandates. Biofuel and biodiesel mandates exist in many countries. For example, in Thailand, the Energy Policy and Planning Office Ministry of Energy introduced domestic mandate of 5% biodiesel blend in diesel fuel, which was later increased to 7%. Similarly, the Royal decree in Belgium sets the binding target for blending biofuels in diesel and petrol (combined) to 8.5% in energy content. Sometimes, the amount/share of mandate may need to be changed due to unexpected events. For example, in Zimbabwe, when the Ethanol Petrol Blending Regulations SI 144 entered into force in 2011, the ethanol blending requirement was 10% ratio for gasoline, two years later it was increased to 15%, but due to low supplies, it was decreased to 5% in 2015 [5]. Due to the obligatory nature of such policies, they tend to guarantee the increased deployment. However, compliance should be ensured via proper management. Also, in case of zero-emission vehicle mandates, the way that the vehicles will be tested for their emissions should be transparent in order to increase trust. Moreover, such testing can increase capital costs due to the necessity of appropriate testing facilities and technical expertise.

Zero emission vehicle mandates can be applied in different forms. For example, according to [5], in United Kingdom, there will be no new sales of new petrol and diesel cars and vans from 2030, although hybrid cars and vans with longer distance capabilities will be allowed to be sold until 2035. Such mandates can also be applied as Low Emission Zones or Zero Emission Zones. As described in [10], low emission zones (LEZs) are usually the areas where the most polluting vehicles cannot enter the area. In some low emission zones, however, the more polluting vehicles have to pay more if they enter the low emission zone. Similarly, in zero emission zones (ZEZs), only zero emission vehicles (such as battery electric or hydrogen fuel cell vehicles) are allowed in. ZEZs can be helpful in improving local air quality while encouraging a switch to zero-emission vehicles, public transport use, cycling and walking. This may significantly reduce the country's emissions in which the ZEZs are applied. An example indication for LEZ and ZEZ can be seen in Fig. 10.2. An interested reader may refer to [11] for an in-depth explanation of Local Zero Emission Zones in London, [12] for Oxford, and to [10] for the low emission zones in Europe.

Unfortunately, the improvement in electric vehicle adoption can not only happen due to the existence of LEZs or ZEZs. In some countries, adoption of LEZs or ZEZs can be hard to achieve due to political, economic or social issues. Therefore, there is also a need for a strong alternative fueling infrastructure. Several countries have enacted policies in order to accomplish that. For example, the Green Business Transport Fund in Denmark helps to deploy charging infrastructure for taxis and trucks, as well as biogas and hydrogen projects. Similarly, the government in France announced the opening of a call for projects to support the deployment of high power charging stations for electric vehicles (EVs). The amount of aid for the charging infrastructure deployment can reach up to 40% of the eligible costs. In addition to the examples provided above, in Slovak Republic, 712

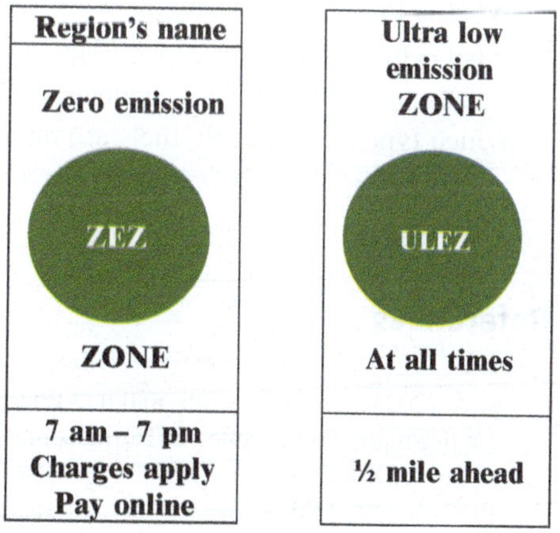

Fig. 10.2 Signage that shows examples of zero emission zone (ZEZ) and ultra low emission zone (ULEZ)

million euros are allocated to support the roll-out of charging stations for alternative fuels, modernization of railways and new cycling infrastructure, based on Slovakia's recovery and resilience plan [5]. Although such policies are really useful for increasing the number of electric vehicles and alternative fuel vehicles in transport, the costs of infrastructure should be balanced very carefully, especially in early stages when the demand is low.

In addition to the policies mentioned above, some countries may apply vehicle standards or vehicle emission standards. Vehicle standard refers to a technical specification that a vehicle structure, system, component or equipment must comply with. According to [13], a vehicle standard is designed to make the road vehicles or vehicle components safe to use, control the gas emissions or noise from road vehicles, secure them against theft, or promote energy saving. Hence, vehicle standards require financial and time investment by original equipment manufacturers. In support of RE and clean energy, the vehicle standards help in reducing tailpipe and evaporative emissions from cars, light-duty trucks, medium-duty vehicles, and some heavy-duty vehicles. There are also some standards, just focusing on emissions, known as vehicle emission standards. Such emission standards define the acceptable limits for exhaust emissions of new vehicles, and thus lead to more efficient use of fuel and lower emissions. Vehicle emission standards are also defined in European Union Directives staging the progressive introduction of increasingly stringent standards. In European Union, different standards apply for each vehicle type. According to [14], noncompliant vehicles cannot be sold in the EU.

10.4 Exercises

1. What is the share of renewable energy in transportation?
2. Which sustainable development goals are endangered by fossil fuel subsidies?
3. Which types of financial incentives can be applied in transportation sector in order to support renewables?
4. What is Power-to-X? What is its relation with RE adoption?
5. What is the main purpose of zero emission zones? In which countries are they applied?
6. What are vehicle emission standards? How do they help RE adoption?
7. Which type of renewable fuels are mandated in transportation sector?

References

1. IEA. (2024). *Transport*. International Energy Agency. Retrieved April 2024, from https://www.iea.org/reports/renewables-2018/transport
2. IRENA, OECD/IEA and REN21. (2018). *Renewable energy policies in a time of transition*. IRENA, OECD/IEA and REN21. Retrieved April 2024, from https://www.irena.org/-/media/Files/IRENA/Agency/Publication/2018/Apr/IRENA_IEA_REN21_Policies_2018.pdf

3. Milicevic, B. (2024). Measuring and monitoring fossil fuel subsidies in the region. Sustainable Energy Division UNECE. Retrieved April 2024, from https://unece.org/sites/default/files/2022-05/D1_1_Measuring%20and%20monitoring%20fossil%20fuel%20subsidies%20in%20the%20region_.pdf
4. United Nations. (2024). *The 17 goals.* Department of Economic and Social Affairs Sustainable Development. Retrieved April 2024, from https://sdgs.un.org/
5. IEA. (2024). *Policies database.* IEA. Retrieved April 2024, from https://www.iea.org/policies
6. CAFRI. (2024). *The importance of decarbonizing New York's transportation sector.* Climate & Applied Forest Research Institute. Retrieved April 2024, from https://cafri-ny.org/wp-content/uploads/2020/02/CAFRIESF-Transportation-Decarbonization-2-Pager.pdf
7. Scott, W. (2017). *Low carbon fuel standards in Canada.* Policy Brief. Smart Prosperity Institute. Retrieved April 2024, from https://institute.smartprosperity.ca/sites/default/files/lowcarbonfuelstandards-web.pdf
8. NRF. (2024). *Renewable fuel standard.* National Retail Federation, NRF. Retrieved April 2024, from https://nrf.com/hill/policy-issues/renewable-fuel-standard
9. Ribeiro, S. K., Figueroa, M. J. N., Creutzig, F., Dubeux, C., Hupe, J., Kobayashi, S., Brettas, L. A. D. M., Thrasher, T., Webb, S., & Zou, J. (2024). *Energy end-use: Transport.* Global Energy Assessment. Retrieved April 2024, from https://previous.iiasa.ac.at/web/home/research/Flagship-Projects/Global-Energy-Assessment/GEA_Chapter9_transport_lowres.pdf
10. Urbanaccessregulations.eu. (2024). *Low emission zones. Urban access regulations in Europe.* EU. Retrieved April 2024, from https://urbanaccessregulations.eu/low-emission-zones-main
11. Mayor of London. (2019). *Zero emission zones: Taking forward the Mayor's transport strategy proposal for zero emission zones.* Transport for London. Retrieved April 2024, from https://content.tfl.gov.uk/tfl-guidance-for-local-zero-emission-zones.pdf
12. Ay Adeduro. (2022). *All you need to know about Oxford's Zero Emission Zone (ZEZ) pilot.* The Oxford Magazine. Retrieved April 2024, from https://theoxfordmagazine.com/all-you-need-to-know-about-oxfords-zero-emission-zone-zez-pilot/
13. Department of Infrastructure, Regional Development and Cities. (2018). *Inquiry into the theft and export of motor vehicles and parts.* Australian Government.
14. Tate, J. (2016). *Vehicle emission measurement and analysis—Aberdeen City Council.* Project report. Institute for Transport Studies, University of Leeds. Retrieved April 2024, from https://www.aberdeencity.gov.uk/sites/default/files/Vehicle_Emission_Tailpipe_Study_Report__April2016.pdf

11 Renewable Energy Policies in Power Sector

Renewable energy deployment in the power sector has expanded dramatically in the last decade, and this trend is expected to continue in the future. According to [1], the share of renewable electricity generation in 2022 was 29.9%, of which 15.1% belongs to hydropower, 12.1% to solar and wind power, and 2.7% to bioenergy and geothermal power. IEA [2] reports that the world will add more renewable capacity in the next five years than has been installed since the first commercial RE power plant was constructed more than 100 years ago. IEA [2] also lists the RE milestones to be achieved as follows:

- 2024: Wind and solar PV (together) generate more electricity than hydropower
- 2025: RE becomes the largest source of electricity by surpassing coal
- 2028: RE sources account for over 42% of global electricity generation

The fast growth in adoption of renewable electricity is due to the fact that majority of countries around the world provide policy support and have national targets. This chapter is devoted to RE power policies.

The RE policy types covered in this chapter are mostly the policy instruments listed in [3] for promoting the use of renewables in power sector. The categorization adopted in this book is also followed in this chapter. Figure 11.1 illustrates the related renewable energy policies according to their categories; incentives, education and research, public finance and regulations.

N. Tekbıyık Ersoy, *Energy Efficiency and Renewable Energy Policies*, Synthesis Lectures on Renewable Energy Technologies, https://doi.org/10.1007/978-3-031-64305-7_11

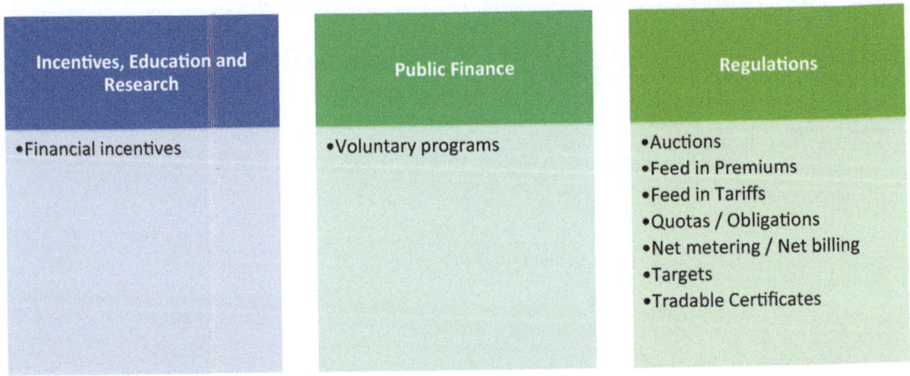

Fig. 11.1 Renewable energy policies adopted in power sector

11.1 Incentives, Education and Research

In power sector, incentives can be applied in the form of grants, tax credits, investment subsidies, etc. As these policies have been explained in detail before, they will not be elaborated here again. It is clear that financial incentives increase the affordability of technology and reduce capital costs. However, support or incentive levels can change frequently depending on the changing political priorities of the countries. For example, according to [1], in 2022, Norway increased the maximum subsidy per kW installed and the maximum system size eligible for rebates, from 15 to 20 kW in order to incentivize residential systems. Similarly, Germany reduced the value-added tax (VAT) to 0% for residential PV systems up to 30 kW and provided tax exemptions to small PV system operators. Although financial incentives are very beneficial, at this stage it should be noted that the incentives focusing on only the installation sizes, rather than the amount of production, do not necessarily motivate the RE investors to produce more electricity after they have received the incentives.

11.2 Public Finance

Voluntary programs applied in power sector have the benefit of enabling the deployment with no involuntary extra cost to the government or the consumers. For example, in United States, the Environmental Protection Agency's Clean Energy Supply Programs are voluntary programs that help in increaseing the use of renewable-sourced power and Combined Heat and Power (CHP) systems among leading U.S. organisations. Programs include: Green Power Partnership and CHP Partnership [4].

In order to increase the amount of voluntary programs, first the level of awareness has to be increased. This can be done with carefully designed awareness programs. A good example to that can be the one applied in Lao People's Democratic Republic (PDR). The Policy on Sustainable Hydropower Development in Lao PDR offers policy guidance to agencies that oversee investment projects in the hydropower sector [4].

11.3 Regulations

There are many different types of regulatory measures used in promoting RE in power sector, such as; auctions, feed in premiums, feed in tariffs, quotas/obligations, net metering/net billing, targets, and tradable certificates. Auctions provide flexibility in design, as they can be tailored according to the needs. Moreover, they allow the discovery of the real price as the possible investors compete in auction. This motivates them to offer the lowest prices for the highest quality projects. According to [1], between January–September 2022, awarded renewable auction capacity increased 70% to reach 77 gigawatts (GW), primarily in solar PV and wind power. Moreover, Philippines announced a Green Energy Auction Programme for 2023 of a combined 2 GW for biomass, hydro, solar and wind power capacity. Similarly, Poland and South Africa awarded auctions for solar power. The German market was also mainly driven by government auctions and more than 3 GW of tenders. Although many auctions/tenders are held around the world each year, REN21 [1] reveals that in 2022, the countries have struggled to adjust to price fluctuations in materials and energy, and this resulted in undersubscribed auctions and tenders. The reason was adhered to their conditions not matching the market reality.

In addition to the auctions, Feed-in tariffs (FITs) and feed-in premiums (FIPs) have been used widely to support renewables within both large-scale grid systems and for decentralized power generation. Typically, auctions are used for large-scale projects and FITs and FIPs are used for small-scale installations. Also, FIT is suitable in markets with low level of RE development. According to [1], by the end of 2022, 83 countries had in place FIT or FIP policies. FIT is a popular mechanism as it offers limited risk for RE developers. As the price is fixed, the FIT beneficiaries are not affected from electricity price volatility. However, when the market price is higher than the tariff level, they cannot benefit from it. In contrast to that, FIPs provide an incentive to produce electricity even when supply is low, as in such situations, usually the market prices are higher and the RE investors benefiting from FIPs can receive higher amount of payment. However, it is hard to set and adjust the tariffs (in FIT) and caps/floors for the premiums (in FIP) when cost structures change dynamically [3]. Hence, FIT and FIP need to continuously be adapted to changing market conditions. This allows to achieve a greater cost-competitiveness and improved RE integration. For example, in 2022 and early 2023, 15 countries revised their feed-in tariff or premium payment (FIP) policies [1]. According to the data provided in [5], fixed tariffs and premiums are predicted to be the primary driver of renewable electricity capacity between 2023 and 2028 with 56%, followed by the auctions with 19%, and tax credits with 11%, etc.

Quotas and obligations are other RE policies applied in power sector. The obligations can be in the form of renewable portfolio standards (RPS), or renewable obligations. According to [1], by the end of 2022, 35 countries had in place RPS, which mandate utilities and private sector companies to install or use renewables. The advantage of obligations is that they can be started with low obligations and then, increased over time. For example, in 2022, Philippines updated its RPS to require a minimum of 2.5% RE supplied to distribution utilities or direct buyers, up from 1% previously. These obligations can also be technology specific. Actually, electricity quota obligations vary in their design according to the choice of design elements. These elements can be the time frame, technology, obligated entities, compliance rules, etc. In case of solar, mandates for rooftop solar PV require certain buildings to install solar panels on their roofs. Even when technology is defined, the obligations may have other restrictions or conditions. For example, as reported in [1], EU requires the installation of rooftop solar PV on all new public and commercial buildings with at least 250 m^2 of surface by 2026, on all existing public and commercial buildings by 2027 and on all residential buildings by 2029. Sometimes, even the building types can also be stated. The new legislation in France makes it mandatory for parking lots of 80 spots or more to install solar PV systems within three to five years.

In power sector, quotas are generally supported by renewable energy certificates (RECs). A REC is usually awarded to a generator for each MWh (Megawatthour) of RE produced. The suppliers or generators receive or buy certificates to meet the mandatory quotas each year. Quotas and obligations, along with tradable RECs, are applicable to all stakeholders for installations of various sizes. However, the effectiveness of them depends on specific market design considerations and how well the non-compliance is monitored/penalized. According to [3], if the amount of penalty is not enough for the entities that fall short of the required number of RECs each year, the certificates will not have value on the market and thus, will not drive greater deployment. If the demand for certificates is to be ensured, the penalties should be higher than the market price of the certificates. Because, if not, the entities may prefer to pay the penalties rather than trying to reach the targets.

The main idea behind the quotas/obligations is to make sure that the RE targets of the country can be achieved. Most countries have in place national RE targets for the power sector. According to [1], as of 2022, 174 countries had targets for renewable power shares (with 37 of those countries' targets being 100% renewable electricity). A country may have medium term or long term targets. For example, Japan set targets for 10 GW of offshore wind capacity by 2030 and 30–45 GW by 2040. Similarly, Germany have targets of reaching 20 GW (Gigawatts) offshore wind energy by 2030, and 40 GW by 2040 [4]. Although targets provide clear signals to the electricity consumers, unfortunately, they are significantly affected from the political situation in a country and the related commitments. This makes targets to be not effective on their own and to require complimentary policy measures for implementation.

Net metering and net billing are appropriate measures to support distributed generation. Distributed generation happens when many RE power plants or prosumers are connected to the grid and they can transfer electricity to the grid. With net metering and net billing, the owners of residential or commercial buildings can have individual installations, such as rooftop PVs. According to [6], the main difference between net metering and net billing lies in the fact that under a net-billing scheme, the difference between injected and consumed electricity is calculated based on the bills for injected and consumed electricity. Two meters are installed in order to generate two separate bills. In net metering however, the injected electricity is counted as 1:1 to consumed electricity and the prosumer pays for the difference. Mainly, in net metering, the prosumers receive credit for any excess electricity they generate and send back to the grid, which can be used to offset future electricity bills. But in net billing, the customers are paid for the excess energy they generate at a fixed rate per kWh, which is usually lower than the retail rate they would pay for electricity [1].

Net metering is a popular policy for incentivizing households, commercial entities and industrial facilities to invest in RE. According to [1], by the end of 2022, a total of 92 countries had net metering policies in place. For example, in Netherlands, nearly half of the new installations in 2022 were rooftop PV, driven largely by the country's net metering policy. Sometimes, the effect of net metering can be much more than expected. For example, Poland became one of the top 10 solar PV installers in 2022 thanks to the net metering scheme, as the residential prosumers represented around 80% of the new capacity. However, when the number of prosumers increase considerably, the high amount of RE injection can cause significant challenges for the power grid. In relation with that, in April 2022 Poland replaced the net metering with net billing scheme as it is a comparably less attractive option for the households [1]. Unfortunately, possible grid congestion is not the only problem related with net metering. Applying net metering may also cause cross-subsidization problems. Also, tariff changes can have a significant effect on a system's payback period.

11.4 Exercises

1. What is the share of renewable energy in electricity?
2. What are the drawbacks of financial incentives focusing on only installed capacity?
3. Which type of regulatory policies are applied in power sector in order to support renewables?
4. What is the main difference between net metering and net billing?
5. What are the drawbacks of adopting RE targets in power sector?
6. Which RE policies are preferred for large-scale projects? Which ones are preferred for small-scale projects?

7. Electricity quota obligations vary in their design according to the choice of design elements. What are those elements?

References

1. REN21. (2023). *Renewables 2023 global status report*. REN21. Retrieved April 2024, from https://www.ren21.net/wp-content/uploads/2019/05/GSR-2023_Energy-Supply-Module.pdf#p13
2. IEA. (2024). *Renewables*. International Energy Agency. Retrieved April 2024, from https://www.iea.org/energy-system/renewables
3. IRENA, OECD/IEA and REN21. (2018). *Renewable energy policies in a time of transition*. IRENA, OECD/IEA and REN21. Retrieved April 2024, from https://www.irena.org/-/media/Files/IRENA/Agency/Publication/2018/Apr/IRENA_IEA_REN21_Policies_2018.pdf
4. IEA. (2024). *Policies database*. IEA. Retrieved April 2024, from https://www.iea.org/policies
5. IEA. (2024). *Electricity*. International Energy Agency. Retrieved April 2024, from https://www.iea.org/reports/renewables-2023/electricity
6. EU. (2024). *Net-metering (support scheme for RES for own consumption 2020-net-billing). Clean energy for EU islands*. European Commission. Retrieved April 2024, from https://clean-energy-islands.ec.europa.eu/countries/cyprus/legal/res-electricity/net-metering-support-scheme-res-own-consumption-2020-net

12 Technology Based Renewable Energy Policy Analysis

12.1 Technology Based Analysis of RE Policies

This section provides a technology based analysis of the adopted global RE policies. According to [1], in 2023, 36.5% of the installed RE capacity belonged to solar energy, 32.8% was hydropower (excluding pumped storage), and 24.4% was wind energy. Considering that hydropower, wind and solar installations constitute an important part of global installed RE capacity, the analyses were conducted for each one of these technologies. This analysis is performed by using International Energy Agency's (IEA) energy policies database [2]. In that database, the following filters have been applied: Topic: Renewable energy, Status: In force. Then, three different keywords have been used for searching among the renewable energy policies; solar, wind, hydro. Results revealed that there were 539 solar energy related policies, 453 wind related policies and 99 hydropower related policies.

The technology specific analysis results show that solar energy receives the highest amount of support, in terms of number of policies used to support the technology. Figure 12.1 shows the number of RE policies used for supporting solar energy in top ten countries, globally. It is clear that India, United States and China are the leaders. This is mostly in line with the fact that as of 2023, the global leaders in terms of installed solar energy capacity are; China, USA, Japan, Germany, and India, respectively [3].

Figure 12.2 shows the number of RE policies used for supporting wind energy in top ten countries, globally. The leaders are the same as the in case of solar, however, with a rank change; United States, China and India, respectively. When the installed wind energy capacity is observed for 2023, it is seen that the leaders are; China, USA, Germany, India, Spain, respectively [3]. Although the leaders of the two comparison cases are not perfectly aligned, it is evident that there is a strong link between wind energy related RE policies and wind energy installations.

N. Tekbıyık Ersoy, *Energy Efficiency and Renewable Energy Policies*, Synthesis Lectures on Renewable Energy Technologies, https://doi.org/10.1007/978-3-031-64305-7_12

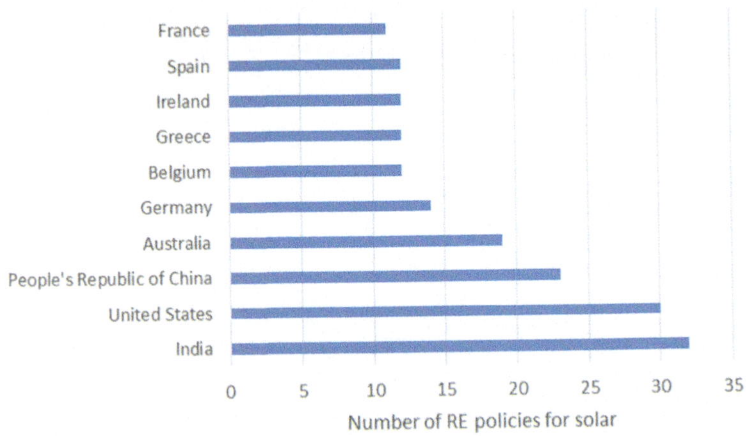

Fig. 12.1 RE policies used for supporting solar energy

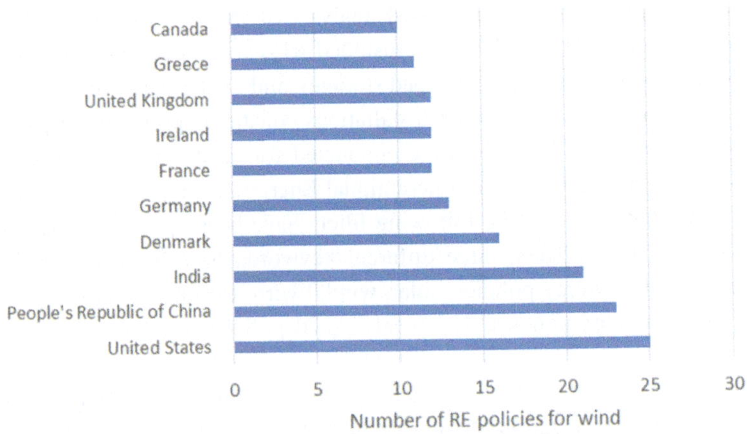

Fig. 12.2 RE policies used for supporting wind energy

When hydropower related policies are analyzed (as seen in Fig. 12.3), the first thing to realize is the amount of policies being much less than the ones offered for other technologies. This can be due to the fact that, hydropower was among the first renewables to be used for electricity production, and therefore many countries with enough water resources have already been investing in hydropower for so long. Hence, they do not require extra support mechanisms. The global leaders in installed hydropower capacity (excluding pumped storage) are [3]; China, Brazil, USA, Canada, and Russian Federation, respectively. When it comes to the amount of RE policies, United States is the

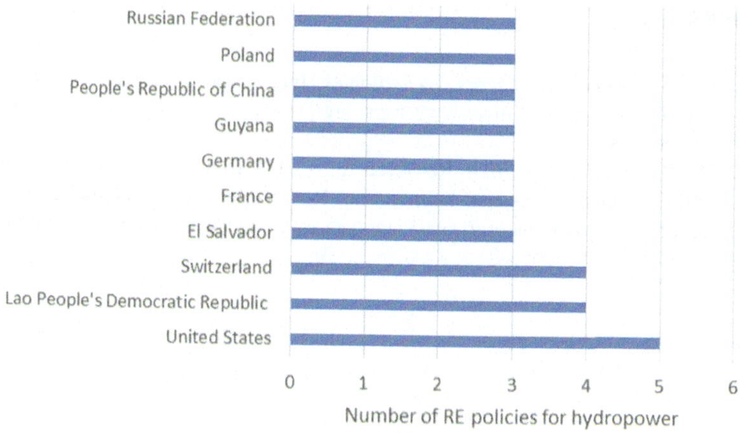

Fig. 12.3 RE policies used for supporting hydropower

leader, followed by Lao People's Democratic Republic and Switzerland, respectively. This mismatch is actually in line with the argument stated above.

12.2 Exercises

1. Which countries are the leaders based on the number of RE policies provided to support solar energy?
2. Which countries are the leaders based on the number of RE policies provided to support wind energy?
3. Which countries are the leaders based on the number of RE policies provided to support hydropower?
4. Why did solar energy receive the highest amount of support, when compared to other technologies?
5. What are the main reasons of leader mismatches when the number of RE policies and installed RE capacity is compared?

References

1. IRENA. (2024). *Renewable energy technologies*. International Renewable Energy Agency. Retrieved April 2024, from https://www.irena.org/Data/View-data-by-topic/Capacity-and-Generation/Technologies
2. IEA. (2024). *Policies database*. IEA. Retrieved April 2024, from https://www.iea.org/policies
3. IRENA. (2024). *Country rankings*. International Renewable Energy Agency. Retrieved April 2024, from https://www.irena.org/Data/View-data-by-topic/Capacity-and-Generation/Country-Rankings

The manufacturer's authorised representative in the EU is Springer
Nature Customer Service Centre GmbH, Europaplatz 3, 69115 Heidelberg,
Germany. If you have any concerns regarding our products, please
contact ProductSafety@springernature.com

Printed and bound by CPI Group (UK) Ltd, Croydon, CR0 4YY
30/09/2026
02225575-0019